T0229847

SEISMIC EFFECTS OF BLASTING IN ROCK

SEISMIC EFFECTS OF BLASTING IN ROCK

A.A. Kuzmenko
V.D. Vorobev
I.I. Denisyuk
A.A. Dauetas

RUSSIAN TRANSLATIONS SERIES
103

A.A. BALKEMA/ROTTERDAM/BROOKFIELD/1993

Translation of : *Seismicheskoe deistvie vzriva v gornikh porodakh*
Nedra, Moscow, 1990

Translator   : Mr. S. Sridhar
Technical Editor : Prof. A.K. Ghose
General Editor   : Ms. Margaret Majithia

ISBN 90 5410 214 4

Distribution in USA and Canada by: A.A. Balkema Publishers, Old Post Road, Brookfield, VT 05036, USA.

Foreword

The development plan of the national economy in our country envisages a significant growth in the mining industry as it is a key industry, satisfying the demand for raw mineral and fuel resources.

The volume of raw minerals and fuel resources mined in the USSR is continuously increasing.

In connection with the intensified production in the mining industry, mainly from open-pit mines, a major role is attached to the utilisation of energy produced by a blast of chemical explosives. It is difficult, as of today, to name a single branch of industry where blast energy is not utilised. Blasting operations are widely conducted not only in the mining industry, but also in agriculture for reclaiming (drying) marshy lands and irrigating arid zones, as also for road laying and special purposes in the oil and gas industry etc. About 2 billion cubic metres of hard rocks are exploited per annum in mining enterprises, while in the irrigation and land reclamation operations, the volume of earthwork considerably exceeds this figure.

The increase in scales and improvement in quality of blasting operations impose rigid requirements on the conduction of industrial blasts to ensure the safety of personnel and protection of engineering structures as well as the surrounding environment. The key measures are directed towards reducing the effect of air shock waves as well as blast-induced seismic effects and decreasing the extent of scatter of rock fragments.

It is possible to develop engineering methods of forecasting and controlling seismic effects in industrial blasts only on the basis of a deep and comprehensive study of the dynamics of wave processes and their relation with natural and technological factors. The effect of blast on rocks is quite complex, just like any other dynamic process that develops in space and time. Its study is associated with considerable difficulties in theoretical as well as experimental investigations and also in the indirect observations under field conditions. Hence many questions concerned with this aspect remain unresolved and inadequately studied, though the theoretical level of knowledge has notably advanced. Of late, the level of practical recommendations for controlling the seismic effects of a blast has reached a higher plane. The implementation of such recommendations has enabled conduction of large-scale blasts (charges weighing up to 1000 t and more) in open-pit mines and also under the congested conditions of industrial and civil structures.

The results of investigations into the seismic effects of blasting charges of various configurations in soft soils and hard rocks are summarised in this book. The problems of controlling blast-induced seismic effects as applied to specific engineering tasks, and also the laws of their variation depending on the various influencing factors (charge configurations, delay interval, relief of slope surface, blasting conditions, type and condition of rocks) were studied.

In the open pits under the Ministry of Road Construction (Mindorstroiy) and the Ministry of Industrial Building Materials (Minpromstroiymaterialov), Ukrainian SSR, the Ministry of Highways (Minavtoshosdor), Lithuaninan SSR and in the metalliferous open pits of 'Soyuzsera' (All-Union Sulphur) group, where operations for compacting subsided land and for constructing structures were undertaken for various purposes under the Irrigation Ministry (Minvodkhoz) USSR, the conclusions drawn from investigations were experimentally verified and the suggested recommendations were implemented very effectively. Such measures enabled avoidance of closure of a number of industrial enterprises, as ordered by the offices of the Government Mining Inspectorate (Gosgortekhnadzhor), and extension of their working lives without incurring capital expenditure on repair and upkeep of the protected targets.

This book is a maiden attempt at generalising the above-indicated results and hence may contain some shortcomings. The authors will gratefully receive any comments and will try to give them due consideration in their future work.

List of Abbreviations/Symbols Used in Text

a —displacement of soil particles, mm;

a_i, b_i —rational numbers, whose sum is unity;

$b = 1/v_1$ —quantity inversely proportional to velocity v_1, $(m/s)^{-1}$;

C —weight of charge, kg ;

$c = 1/v_2$ —quantity inversely proportional to velocity v_2, $(m/s)^{-1}$;

C_1 —charge weight per unit length (linear weight of charge), kg/m;

d_w —width of screening barrier, m;

E_i —seismic energy, J;

E_S —energy flux of spectrum, J;

$F(\omega)$ —amplitude-frequency spectra of displacement velocities;

f —frequency of seismic waves, Hz;

$f(t), f(\tau/r)$ —function of variables t, τ/r;

H —depth at which a charge is placed, m;

$H_{rs} = H/C^{1/3}$ —reduced (scaled) depth of spherical charge, $m/kg^{1/3}$;

$H_{rc} = H/C^{1/3}$ —reduced (scaled) depth of cylindrical charge in a zone away from its effect, $m/kg^{1/3}$;

h_b —depth of barricade, m;

$i = \sqrt{-1}$ —imaginary number;

k_a —anisotropy coefficient;

k_s —coefficient that takes into account the properties of soil;

k —experimental coefficient;

k_σ —experimental coefficient depending on non-dimensional distance;

l_c —length of charge, m;

N_1 and N_2 —distribution of energy according to frequency in the corresponding blasting schemes;

l_b —length of barricade, m;

l_{bh} —length of borehole, m;

m —number of groups being blasted (number of charges);

P — longitudinal seismic wave;

p — complex variable;

$p(t)$ — pressure at the wave front, Pa;

R — direct seismic wave (R_1, R_2 — amplitudes of direct wave);

r — distance from place of blasting to observation point, m;

r_c — radius of charge, m;

r_e — radius of seismic wave emission (source) of a blast, m;

r_s — seismically safe distance, m;

S — transverse seismic wave;

$S(\omega)$ — amplitude-frequency spectra of wave;

$S(j\omega)$ — complex amplitude-frequency spectra of wave;

T — oscillation period of seismic wave, s;

T_0 — period of natural oscillations, s;

t — current period of time, s;

t_0, t_{max} — time periods corresponding to the beginning of perturbance and its growth to the maximum extent, s;

u — Velocity of soil particles, cm/s;

u_P, u_S — velocity of displacement in longitudinal and transverse waves respectively, cm/s;

u_{per} — permissible velocity of waves, cm/s;

v — velocity of wave propagation, m/s;

v_P, v_S — propagation velocity of longitudinal and transverse waves respectively, m/s;

w_v — moisture content in soil;

β — coefficient characterising duration of positive phase of stresses;

β_i — coefficient of dampening of displacement velocity;

R_1 — phase of ground wave;

$\Delta\tau$ — delay interval, s;

ε — volumetric deformation of medium;

η — screening coefficient;

η_t — reduction coefficient of seismic effect during short delay blasting of shot holes (SDB);

θ — coefficient characterising steepness of accrual and drop of stresses;

λ, μ — Lame's constant;

λ_w — wavelength, m;

μ_σ — index of stress attenuation;

ν — Poisson's coefficient;

n — index of level of attenuation maximum of displacements (rate of displacements) over distance;

ξ — ratio v_S/v_P

ρ — rock density, kg/m^3;

σ — ultimate compressive strength, Pa;

τ — total duration of blasting effect, s;

τ_1, τ_2 — time of accrual and drop of stresses respectively, s;

ψ — ratio v_R/v_S

ω — angular frequency, rad/s;

ω_0 — natural angular frequency, rad/s.

Contents

1

Fundamental Concepts on Rocks and Seismic Effects of a Blast

1.1 Characteristics of Rocks

Given the current industrial blasting practice, it is necessary that structures built on soil as well as rocky foundations be protected. This stipulates stringent requirements for the selection of blasting parameters from the point of view of seismic safety, to protect the industrial and civil structures. In selecting such parameters, the first consideration must be the properties of the rocks to be blasted.

The objectives of (blast) effect determine the volumes (extent) of rocks in whose limits it is essential to study their specific properties.

It is difficult to find in different deposits, even two rocks of similar composition which have the same properties. Therefore, in each specific case, knowledge of the characteristic response of rocks subjected to blasting acquires a greater significance.

The physicomechanical properties of rocks, the effect of mineral composition and structure of rocks on the variation in such properties and also the effect of physical fields on rocks have already been adequately discussed [32]. The principal physical properties of rocks found in different deposits of the USSR have also been systematically classified [33].

The magnitude of seismic effects on the rock massif is determined by structural-textural features of rocks (stratification, cleavage, joints), composition (mineral) and the bonding nature between individual grains.

Specialists studying rocks classify them according to the methods and indexes related to their major field of activity. Thus each field of specialisation pursues its own objectives: geophysics — study of the structure and history of development of the planet; geology — establishing the laws of rock formation and subsequent changes in them; exploration — locating deposits, estimating reserves and determining their characteristics; mining — extracting minerals and supporting the mined-out area; civil — providing stability to structures etc.

Usually rock properties can be judged by the ratio of intensity of the blast effect to that of the actual effect (for example, ratio of work done to the degree of rock fragmentation characterises rock hardness). Thus, the determination of rock properties is selective.

Rocks are classified according to various indices. The properties are interlinked and also affect each other. From among the known classifications, mining and civil engineers adopt those which have more common properties bearing on the technological processes. Such classifications are mainly preferred which offer quantitative measures of rock properties for simple and convenient usage in engineering calculations of technological process parameters. Some examples are: rock classification based on hardness (M.M. Protod'yakonov), drillability, stability, plasticity, Young's modulus, blastability etc. Each classification has its own merits and demerits.

Without indicating any specific classification of rocks based on other well-known characteristics, let us consider only those based on the nature of bonding between individual grains [32]. Such a classification is of much greater importance in blasting practice. Rocks are grouped into: loose (distinctly granular) rocks, in which there are no links between minerals or grains (sand, gravel, pebbles); bonded (clayey) rocks with a water-colloidal bond between particles (clays, loam, bauxites); and hard (very hard and hard) rocks with rigid and elastic links between mineral particles (granites, limestone, diabase, gneiss, marble etc.).

The distinctive feature of rocks is their polyclustering nature, as the pores and fissures in natural conditions are filled with gases, liquid (water) or other rocks, which result in a wide spectrum of physicomechanical properties.

The more pronounced properties of rocks, grouped according to the above-mentioned parameters, are mentioned briefly in the following table: soft soils, frozen soils and hard rocks. The patterns of seismic wave propagation in such rocks during blasting have their own distinguishing features.

Soft Soils: Blasting operations in certain cases are required to ensure the safety of different structures from damage due to seismic effects. Structures built on a soft soil foundation exert heavier loads compared to those built on a rocky foundation.

Soils are multiphase media with a dispersed mineral structure and continuous variation in physicomechanical properties. Soils are a set of mineral particles (grains) with mutual contact, but not filling the entire space. There is no cementation between grains in soils because of their porous structure and the solid particles form the skeleton. The voids are filled with liquid (water) and gaseous (air, water vapours, carbon dioxide) phases existing in both free and associated conditions. Water may be present in a solid form (ice) which materially changes the physicomechanical properties of soils. In a completely water-saturated soil, where gas is not present, the soil becomes a two-phase system. Soils which are not entirely saturated constitute a three-phase system. Most such soils consist of solid particles with voids in-between.

The mineralogical composition of solid particles, their shape and dimensions exert considerable influence on the properties of soils. The quantitative relation between mineral grains of different sizes is described by the granu-

Table 1

Type of particles	Size of particles, mm	Individual particles	Size of particles, mm
Boulders rounded, rocks, angular	>200	Large Medium Small	800 800–400 400–200
Rounded pebbles, crushed stone, angular	200–40	Coarse stone Crushed stone, lumps of pebble Small rubble, small pebbles	200–100 100–60 60–40
Rounded gravel	40–2	Coarse Medium Small Very small	40–20 20–10 10–4 4–2
Sands	2–0.05	Coarse Big Medium Small Fine	2–1 1–0.5 0.5–0.25 0.25–0.1 0.1–0.05
Dust	0.05–0.001	—	—
Clays	<0.001	—	—

lometric composition of soils. The classification of soils as per granulometric composition (after V.V. Okhotin) is given in Table 1.

The physicomechanical and dynamic properties of soft soils vary over a wide range depending on the size of particles and composition of components. Longitudinal and transverse waves exert considerable seismic effects on soils. The velocity of propagation of these waves depends on the mechanical (elastic) properties of soils and density (Table 2). The higher the density and elasticity of soils, the greater the velocity of propagation of longitudinal waves. Knowledge of the laws of propagation of seismic waves through soft soils is very important in blasting operations since most of the structures to be protected in mining enterprises and civil projects are built mostly in or on such soils.

The analytical and experimental relationships of spherical and plane wave propagation in water-saturated and unsaturated rocks are given in many published works. In such publications, the results of investigations were applied to various practical problems associated with the geodynamics of blasting operations [24].

The specific values attributed to the characteristics of seismic waves are related to soil conditions. This may be expressed quantitatively and considered in practical recommendations for any technological process. However, it is to be noted that in actual conditions very few instances are observed in which there is a change in only one of the characteristics of soil conditions. Usually, the

Table 2

Rocks	Density g/cm³	Velocity of elastic waves, km/s		Seismic rigidity
		Longitudinal	Transverse	
Sands of different grains, pure:				
naturally moist	1.4–1.6	0.2–1	0.1–0.7	0.3–1.6
water-bearing	1.85–2.15	1.5–1.8	—	2.8–3.7
aerated (loose)	1.3–1.4	0.1–0.4	0.04–0.3	0.06–0.9
Sands mixed with clayey material (up to 5%):				
naturally moist	1.4–1.6	0.3–0.8	0.1–0.6	0.4–1.3
water-bearing	1.8–2.1	1.5–1.75	—	2.7–3.7
Loamy sand:				
naturally moist	1.45–1.9	0.3–0.7	0.1–0.35	0.44–1.3
water-bearing	1.8–2	1.7–1.9	—	2.8–3.8
Loam:				
naturally moist	1.65–2.05	0.3–0.9	0.08–0.45	0.5–1.8
water-saturated	1.7–2	1.6–1.9	—	2.8–4
Clays:				
Naturally moist	1.3–2	0.85–1.4	0.2–0.7	1.4–2.8
water-saturated	1.8–3.25	1.75–2.2	—	3.1–7.1
Loess loam and yellow-grey loam:				
naturally moist	1.16–1.75	0.3–1	0.1–0.7	0.5–2.5
water-saturated	1.6–2.6	0.15–0.5	0.02–0.08	0.2–1.3
water-bearing	1.6–2.6	1.5–1.8	0.1–0.7	2.4–4.7
Loose:				
unsaturated	1.3–1.5	0.03–0.3	0.01–0.2	0.04–0.5
water-saturated	1.5–1.8	1.5–1.7	—	2.2–3
Soils	1.4–1.85	0.04–0.5	0.01–0.2	0.06–0.9

engineering-technological indexes and seismic parameters vary under the influence of several factors. It is very difficult to distinguish the mutual relationship of any two quantities by themselves [25]. In such a case, it is advisable to plan the experiment such that it is possible to monitor the mutual relationship of the parameters that are of interest and that too particularly for a specific mining condition.

Frozen Soils: Such soils exist at zero or subzero temperatures and at least a part of the enclosed water is in a frozen state. Frozen soils are categorised as a four-phase system, consisting of solid mineral particles, viscous-plastic inclusions of ice, water in closely and loosely held states and gaseous components (water vapours and gas).

Depending on the duration of the state of existence, such soils are classified into permafrost (several centuries and several thousands of years), frozen for long periods (from several years to several decades), seasonally frozen (one or two seasons) and frozen for short periods (from several hours to several days).

Permafrost soils cover 25% of the dry land of the earth. In the northern and north-eastern parts of the USSR they occupy about 11 million km^2 (about 49% of the territory of the USSR). The rest of the territory is covered with seasonally frozen soils [15].

The thickness of the permafrost soils and those frozen for long periods is not constant in different regions of the country. For example, in the Chitinsk region it ranges up to 20 m, in the Yakutsk Autonomous Republic up to 230 m, in Novaya Zemlya up to 600 m and in the Vilyui River area up to 800 m. The temperature of such soils also varies over a wide range: in Vorkut from -0.3 to $-1.5°C$, in Yakutsk from -3 to $-7°C$, in Novaya Zemlya up to $-12°C$ and in Pevek around $-5.2°C$ [15].

The surface layer thickness of frozen soils, freezing or thawing depending on the season, is usually termed an active layer. The properties of soils found in this layer may differ from those of bedded soils. The thickness of the active layer of perennially frozen soils varies from 0.7 to 3 m.

The physical condition, structure and properties of both frozen and thawed soils depend on the type of water held by them. The lower the dispersion of the soil, the stronger the effect of water on the properties and the more complex the nature of interaction of water with its components. The type of water determines the nature of migration, ice formation, soil heaving and other processes during the frozen state of the soil. Water may exist in different phases in the soil: vapour, liquid and ice. The nature of the bond between water and the soil skeleton is also quite distinct.

Presently, the following types of water in soils are distinguished: vapour, hygroscopic, film water, loosely held, free (gravitational, capillary, closed capillaries) and ice.

Film water is at a temperature lower than $-1.5°C$. Its maximum content, expressed in a fraction or percentage of the weight of dry soil, is termed the *maximum molecular moisture capacity*. This index depends on the chemical, mineralogical and granulometric composition of the porous skeleton. Soils are classified in the following manner (G.P. Mazurov as per the maximum molecular moisture (%).

Clay	24
Loams:	
heavy	24–16
medium	16–12
light	12–8

Loamy sands:

heavy	8–4
light and fine sand	< 4

Frozen soils are grouped according to granulometric composition based on the content of clayey particles in them. The ratio by weight of clayey particles to that of hard particles is as follows: clay, more than 30%, loam 30–10%, loamy sand 10–3%, fine sand, less than 3%.

Frozen soils are highly sensitive to changes in the external environment: temperature, pressure, duration of freezing, rate of loading etc. The main principle of frozen soil mechanics establishes the dynamic balance in such soils between values of external effects and the amount of unfrozen water and ice. The changes in physicomechanical properties of frozen soils due to variation in external effects is explained on the basis of this principle.

The principal physicomechanical properties of soils are: density, total mass density, density of solid mineral particles, moisture due to unfrozen water or relative degree of icing, compressive strength, shear strength etc. Other properties are expressed through the main indexes, for example the density of soil skeleton, coefficient of porosity, total moisture content, extent of freezing in all directions, elasticity modulus, shear modulus etc. Some physical indexes of frozen soils have been tabulated (Table 3) [15]. The strength characteristics of such soils are dependent on the quantitative ratio of the indexes of their properties.

It has been established [15] that the intensity of seismic vibrations due to blasting is similar in both frozen soils and hard rocks. The seismic effect of a blast is reduced during the transition of a frozen soil to a thawed soil. Therefore, it is essential to know the properties of frozen soils, particularly their structural features, so that protection may be guaranteed to the buildings erected on such soils.

Hard Rocks: The majority of igneous and metamorphic rocks and some sedimentary rocks belong to this group. The ultimate uniaxial compressive strength of this class of rocks in a water-saturated condition (up to 5%) is about $(5–35) \times 10^7$ Pa. Each rock has a relatively constant mineral content, well-defined chemical composition and also characteristic structure and texture.

Marginal differences in the above features could lead to considerable variations in the physicomechanical properties of even the same rock. Rocks in mineral deposits are subjected to complex stress states and they are usually fissured. The natural condition of rocks is characterised by all these features in toto. They require to be studied for executing technological processes.

Rock massifs happen to be complex discontinuous media, divided into separate structural blocks of various sizes. Such massifs are distinguished by the type of fracturing, degree and nature of anisotropy and non-homogeneity. The variation in rock properties in individual structural elements, governed by general

Table 3

Type of soils	Density, t/m³		Moisture content, vol. %	Porosity
	Soil	Skeleton		
Gravelly-pebbly	2.2	1.95	25	0.38
Coarse-grained sands	2.1	1.85	30	0.46
Medium-size sands	2.05	1.75	35	0.54
Fine-grained sands	1.9	1.5	42	0.8
Loamy sands	1.8	1.4	47	0.93
Loams	1.7	1.2	58	1.25
Crushed stone mixed with wood fillings and clays	1.95	1.65	37	0.65

non-homogeneity of the massif, is of even greater significance. The properties of the massif and its natural components are not similar. In the process of break-age of rocks by blasting, rock properties are determined mainly by individual rock-forming minerals while within the massif they are determined by joints (bedding planes).

One of the main characteristics of hard rocks is density (t/m³), the values of which are given below for some selected rocks [32].

Igneous Rocks

Basalt	2.75–3.2
Gabbro	2.75–3.15
Granite and dionite	2.6–2.8

Sedimentary Rocks

Dolomite	2.6–3.2
Limestone	2.4–3
Rock salt	2.12–2.22

Metamorphic Rocks

Ferruginous quartzite	2.8–4
Talc	2.7–2.8
Quartzite	2.65–2.7
Marble	2.6–2.7
Serpentinite	2.5–2.65
Argillite	2.06–2.7
Chalky clay (marl)	1.84–2.74
Aleuvrolite	1.75–2.97
Sandstone	1.53–2.95

| Anthracite | 1.4–2 |
| Coal | 0.8–1.25 |

It is possible to specify the type of drilling, loading and crushing equipment, once we have the data on the density of hard rocks. Further, the control of the blasting processes within assigned limits also becomes possible.

Among the parameters of joints one should know and consider the type of joint matrices, angles of dip and azimuths of the main joint systems, their aperture and depth, nature and extent of filling, geometry of blocks in the massif etc. The predominant parameters should be considered while studying the blasting effect. The nature of seismic wave propagation in a blast is greatly influenced by the joint system in the massif and its relationship with the non-homogeneity of the massif. This relationship and its effect on blasting have already been adequately discussed [6].

It is to be noted that there is no unanimity of view in describing theoretically the dynamic effects of a blast on a rock massif nor in forecasting the state of the massif after the effect is over [29].

Several theories of rock breakage by blasting are well known (E.G. Baranov, A.F. Sukhanov, O.E. Vlasov, G.I. Pokrovskii, A.N. Khanukaev and others), which are based on general principles of physics and mechanics of continuous media. These hypotheses incorporate mechanical theories of strength–failure criteria of Mises, Coulomb-Mohr's theory, crack Griffith's theory etc.

By assessing the physical effect of certain phenomena on the medium, rock can be represented as a homogeneous continuous medium having some averaged parameters. The parameters are averaged by approximating several sets of experimental data and obtaining empirical relationships and coefficients. When we make assumptions and simplifications in representing rock as a homogeneous continuous medium, it becomes impossible to uniquely evaluate the rock massif and to select a unique physicomathematical model of the phenomena and processes occurring. It is advisable, therefore, to consider the rock massif as a mechanical system, i.e., a system of masses located in space having mechanical links between them. This requires establishing the mechanical bonds between masses and grouping macrojoints according to the type and properties of mechanical linkages between individual particles. Here three types of joints can be distinguished — closed joints which have all the three types of mechanical links (elastic, viscous and plastic); closely held, having only two types of links (viscous and plastic) and open joints with no links.

It is advisable to use structural rheological models of elastic, viscous and plastic elements of bonds [29] to describe different properties over time as well as the properties of mutual links between two discrete rocks in the massif.

The criteria of maximal values of normal and tangential stresses should be upheld so as to distinguish in any given classification between the microstructure and macrofractures and also to determine weight of the mechanical model. Mechanical properties of rock blocks are identical to mechanical properties of the actual rock. Here it is convenient to classify rock massifs according to jointing along three projections of the linear dimensions of the blocks with their sequence in space as per the type and form of joints and their condition at the time of blast effect. Then it becomes possible to divide the rock massif into masses of different sizes depending upon the objectives and possibilities of investigations as required by the formulation of the practical problem.

A relative classification is more suitable for simplifying the calculations and for assessing realistically the processes or phenomena occurring in the rock massif or in a limited volume of a block of rocks [29].

If the blasting of explosive charges is to be considered as a mechanical effect on the massif (shock wave, vibrations, displacements, chipping effects etc.) in a particular sequence, then it so happens that the dimensions of blocks making up this massif are correlated to the wavelengths at the initial period of time, mobility and time period of oscillations etc. As a result, the natural blocks and joints in the massif are subjected to uniform or non-uniform blasting effect in all directions. Then a particular type and form of joints can be transformed into another type and form or can totally transcend the permissible region of approximation.

This concept permits us to construct a mechanical model in any scale for a real rock massif. To develop such a model, it is necessary to distinctly define the masses (blocks) and spatial mechanical links between them by mathematical descriptors. The reliability of the final results is dependent on the accuracy of mathematical calculations as well as the accuracy in establishing the parameters of the actual massif. By using a relative classification, blasting processes in a massif can be described in a simplified manner by selecting a suitable scale for the specific problem.

Objective Evaluation of Hard Rocks (written in collaboration with A.I. Kondrat'ev): It is well established that the seismic effect of a blast in hard rocks is felt differently in different directions [1, 3, 4].

The task of determining the rational parameters which contribute to the seismic effects of a blast arise at the design stage. The determination is mostly prognostic in nature, wherein the reliability is dependent on the level of knowledge about technical-mining conditions and the qualifications profile of specialists. If the dependence of the intensity of seismic effects on charge mass and distance to the object is adequately studied and if it is reflected in the methodologies for calculating blasting parameters, then in practical situations the influence of geological and physicomechanical properties of rocks, structure and construction of massif need not be considered.

Prediction of the seismic effect of a blast is essential at the design stage of a mine as well as for calculating the blast parameters during exploitation. It is necessary to evaluate the probable values of velocities of seismic vibrations in the regions to be safeguarded.

The effect of seismic waves is determined by the nature of wave propagation in the rock massif. It is assumed that at short distances from the source of blasting, the compressive-tensile longitudinal wave exerts the maximum effect. In a zone farther away from the blast, the surface waves exert a greater effect. The intensity of damping of seismic waves depends on the properties of the medium in which they propagate. Elastic waves move with lower losses in hard rocks and are transmitted over larger distances. In loose rocks they dissipate their energy faster and become significantly attenuated at distances away from the blasting site.

Various concepts concerning the impact of the properties of the massif on seismic effects of a blast are based on the analysis of the laws relating elastic and attenuation properties of rocks with their lithological composition, porosity, properties of fluids filling voids and joints, heterogeneity and other geological features. It is well known that as the percentage share of energy expended on breaking rock increases, the seismic effect is reduced in the near-field zone and vice versa. Hence, it is essential to elaborate those characteristics of the massif which influence the seismic effect in large blasts. The means and methods of varying parameters that characterise the properties and structure of a massif and scientific concepts regarding the mechanism of blast effects enable us to upgrade the quality level of design, planning and the technology of the blasting operation itself. An obvious method is to blast each characteristic zone of massif, for which the parameters are calculated in accordance with the structure of the massif and according to fragmentation requirements. In order to solve a specific problem, the characteristics of the massif should be differentially evaluated. Such characteristics are modularity (blocky nature) and joint system. These are the main and physically independent factors. The blocky nature can be represented by the size, form and orientation of blocks in the massif.

Classification of Blocks (Rock Units) in a Hard Rock Massif and Scheme for Determining Their Category

Size	Fine	Standard	Coarse	Boulder
d	$0 < d_f < d_s^l$	$d_s^l \leq d_s \leq d_s^u$	$d_s^u < d_{co} \leq d_{co}^u$	$d_{co}^u < d_b$
Code	1	2	3	4

Transition from class 3 to class 1

Shape	Symmetric	Uniaxially-asymmetric	Asymmetric
	$d^I \approx d^{II} \approx d^{III}$	$d^I \approx d^{II} \neq d^{III}$	$d^I \neq d^{II} \neq d^{III}$
	(difference < 50%)	(difference > 50%)	(difference > 50%)
Code	1	2	3

Transition to class 2

Orientation	Planar	Edge-wise	Peak-wise
α	$\alpha_1 \leq 30°$	$\alpha_{1,\,2} > 30°$	$\alpha_{1,\,2,\,3} > 30°$
Code	1	2	3

Result—category B 312

The following notations are used:

d_f, d_s, d_{co}, d_b	— block size corresponding to the class lump size (f—fine; s—standard; co—coarse; b—boulder)
d_s^l, d_s^u	— lower and upper limits of standard class dimensions
d_{co}^u	— permissible (upper) limit of coarse class
d^I, d^{II}, d^{III}	— sizes of blocks between opposite edges in different (I, II, III) measurements; units of measurement are the same
α_1, $\alpha_{1,\,2}$, $\alpha_{1,\,2,\,3}$	— angle between shock wavefront and single, double, triple edges of block adjoining it.

Fissuring can be specified by the size of aperture, spacing between fissures in each system, azimuth and dip of the fracture plane.

Classification of Fissures in a Hard Rock Massif and Scheme for Determining Their Category

Extent of opening of fissures	Rigid contact	Flexible contact	Absence of contact
Code	1	2	3

Transition to class 2

Nature of fissuring	Highly fissured	Moderately fissured	Weakly fissured	Almost unfissured (monoliths)
ν	$\nu_{hf} > \dfrac{1}{d_s^l}$	$\dfrac{1}{d_s^l} \geq \nu_{av} \geq \dfrac{1}{d_s^u}$	$\dfrac{1}{d_s^u} > \nu_{wf} \geq \dfrac{1}{d_{co}^u}$	$\nu_m < \dfrac{1}{d_{co}^u}$
Code	1	2	3	4

Transition to α_T

Azimuth α_T .... α_T = 0 to 180°

Transition to α_{da}

Dip angle of fissure plane α_{da} .... α_{da} = 0 to 90°

Result—Category T_p 2; 2; 120°; 90°

The following notations are used:

ν_{hf}, ν, ν_{av}, ν_{wf}, ν_m	— descriptor of fissuring in the respective class;
d_s^l, d_s^u, d_{co}^u	— spacing between fissures, corresponding to sizes assumed for assessing blockiness;

α_T — strike angle;

α_{da} — dip angle of fissure plane.

Categorisation of blocks per size is established from the specific requirements and production conditions, while considering the universality of parametric values used in calculations. By production conditions is meant the size of standard classes of blocks (fragments) that limit the dimensions of mechanisms and machines of the technological chain for processing blasted rock. Therefore, blocks are grouped according to size into at least four classes — fine, standard, coarse and boulders.

The shape of blocks can be classified in terms of the relationship of mutually perpendicular linear dimensions between opposite edges (walls). It is advisable to consider the block orientation with reference to the shock wavefront. Three classes can be distinguished — planar orientation, orientation by edges (ribs) or by peak (vertex). Elements with larger dimensions (plane, edges) are preferred as orientation in the case of asymmetric blocks. We speak of orientation of blocks by plane when the plane is at an inclination of not more than 30° to the shock wavefront. Otherwise, the edge-wise orientation can be adopted or if all edges are inclined to the shock wavefront at an angle or more than 30°, then vertex-wise orientation can be adopted.

Such a classification enables one to relate a block to anyone of 36 categories ($4 \times 3 \times 3$). The massif can be represented as containing blocks of several categories. A massif of homogeneous structure consisting of a large number of categories of blocks is highly improbable. They can be established statistically according to the modal distribution.

The aforementioned assessment of a massif according to its blocky nature ought to orient the specialist while calculating the blast parameters towards selecting a suitable scheme for blast initiation, specific explosive consumption and type of explosive etc. The quality of blasting can be enhanced by choosing zones of similar structure in a massif. In practice, a massif can be differentiated within the boundaries of the block to be blasted since variation in a borehole grid permits one to 'tie up' the blast parameters with the selected zones. Zones can be distinguished by several estimated indices: number of categories of blocks contained in a massif; variation in the parameters of blocks; average parametric values of blocks etc.

For a convenient symbolic description of categories of blocks, classes of parameters are numbered according to the size of lumps: d — fine (1), standard (2), coarse (3), blocky (4); as per shape φ — symmetric (1), uniaxially symmetric (2), asymmetric (3); as per orientation α_b — planar (1), edge-wise (2), vertex-wise (3). Then the category in conformity with the coding rules will be of the type: $Bd\ \varphi\ \alpha_b$. Such a coding is convenient in identifying the category visually and can further be used in computerised calculations. For example, B 123 is a block of rock fines uniaxially asymmetric and oriented vertex-wise.

The parameters of fissure are grouped relative to the absorption of blast energy as well as by the orientation of the crushing zones.

Three classes are distinguished according to the extent of fissure apertures: with rigid contacts between blocks; with flexible contacts and with almost no contacts in the natural state of the massif.

To determine the number of fissures between charges, data on their spacing is required. A massif is evaluated by the fissure index ν_T based on each system; evaluation of fissure aperture and orientation is similar. The values of fissure indices are grouped according to the size class of blocks: highly fissured $\nu_{hf} > \dfrac{1}{d_s^l}$; fissured $\dfrac{1}{d_s^l} \geq \nu_T \geq \dfrac{1}{d_s^u}$; moderately fissured $\dfrac{1}{d_s^u} > \nu_{av} \geq \dfrac{1}{d_{co}^u}$; weakly fissured $\nu_{wf} < \dfrac{1}{d_{co}^u}$;

where d_s^l — lower size limit of standard class of unit blocks;

$\quad\quad d_s^u$ — upper size limit of standard class of unit blocks, m;

$\quad\quad d_{co}^u$ — upper size limit of coarse class, m.

The direction of breakage is dictated by the strike of the joints. It is determined by known methods. While designing blast parameters, the strike azimuth is used.

The dip angle of any fissure plane should govern the charge design, angle of inclination of drill holes, type of explosives and point of initiation. The values of the angle of dip should be known at the design stage.

Thus the massif can be characterised by fissures belonging to any one of 12 (3×4) classes and the strike azimuth α_T and dip angles α_{da}.

The suggested principle of classification is universal and applicable to different enterprises and varying conditions prevailing in the same mine. This principle enables one to evaluate the massif, fragmentation requirements and effectiveness of detonation method with the help of scaled parameters determined by the technoeconomic indexes of production. Separate evaluation of a massif as per blockiness and fracturing allows the use of a flexible method for calculating blast parameters by updating it with modern scientific advances in the areas of effective rock breakage by blasting and specific treatment of seismic effects of a blast.

REFINING THE SEISMIC MICROREGIONALISATION OF ROCK MASSIFS

The foregoing assessment of rocks based on modularity and fissuring is one of the constituent elements in the method of seismic microregionalisation in exploiting mineral deposits. This is a formidable task for those mining-technical conditions wherein the buildings to be protected are close to the site of blasting (usually, not less than 200 m). In such a case, selecting the permissible weight of explosive according to the size of the entire pit becomes unsuitable; it is imperative that this parameter be selected for specifically identified zones within the pit. These zones are differentiated by the magnitude of ground motion due to seismic effects. Each zone differs from the other by the extent of modularity and fissuring, physicomechanical properties of rocks and top soil, location relative to structures to be protected, blast parameters and the dynamic characteristics governing the blast. The safe radii as per seismic effect of the blast are determined for each zone in the pit and for the given charge weights (per delay).

Seismic microregionalisation in pit conditions is based on the general principle of establishing zones to justify a set of blast parameters to obtain the desired fragmentation. The distinctive feature of congested conditions in open-pit mines is a constraint imposed by safety regulations for blasting seismically safe charge weights so as to ensure protection of industrial and civilian buildings.

As perceived in practice, the blast parameters should consider the peculiarities of structure of the massif within the zone apart from its physicomechanical properties. The peculiarities are determined by statistically processing data of full-scale measurements of joint elements (strike azimuth, dip angle) and spacing between joints in exposed (slopes) benches of a pit. The mean size of natural rock units in the massif and boundaries of zones based on classes of blocks are established. Later, all the data are entered in the pit plan and subsequently used in planning of blasting operations.

It should be noted that currently while zoning the rock massif it is not possible to fix exactly the boundary between zones nor to forecast their extension at depths for subsequent stages of working the pit field. This is a major disadvantage in the method of full-scale measurements in exposed slopes while zoning open pits and complicates the differentiation of blast parameters. Inevitable errors in establishing zonal boundaries of blocks lead to unsatisfactory fragmentation of rocks and consequently to adverse seismic effects of the blast.

An analysis of the structural location of the series of dykes or dyke-like bodies that develop in pits or form at the boundaries of blocks of various grades as a result of vertical and horizontal tectonic displacements of blocks along conjugate cleavages, plays an important role in establishing boundaries between zones. The presence of dykes is quite characteristic in many types of rocks.

The authors suggest a geophysical method of micromagnetic survey of block profiles in a massif, which facilitates distinguishing rock blocks according to magnetic properties and in fixing the zonal boundaries in blocks. This is an improved method for regionalising open pits. The physical foundation for this method is the dependence of the magnetic induction observed at the exposed surface of the massif on magnetisation and magnetic susceptibility of rocks as well as the extent of discontinuities.

The suggested method was tried in the granite pit 'Selishe-2' (Rovenskii region) using a proton magnetometer MMP-203. Measurements were taken in association with N.N. Shatalov and A.I. Kondrat'ev primarily across the strike of the dykes in three profiles (routes) of observations at +194, +183 and +170 m levels. Observations were taken at an average interval of 3–5 m and sometimes at 1 m. Along each profile, geologic-geophysical sections were drawn along with the mapping of magnetic induction curves $T_a = f(l)$, where

T_a — anomalous vertical component of the magnetic field, nT;

l — length of profile being measured, m.

According to these data, dykes of basic rocks could be identified. By comparing the data from full-scale observations in the pit above the dykes and the granites enclosing them and also by analysing the nature of changes in the curve $T_a = f(l)$, the zones of blocks were determined and their boundaries were fixed, based on the categories and classes (local classification of granites).

Category of blockiness	I	II	III	IV	V
Class of blockiness	< 1.5	1.5–1.9	1.9–2.15	2.15–2.5	> 2.5

It is necessary to note that the elements of deposition of dykes conform,with sufficient accuracy, to the elements of deposition of large joint systems in the massif of bedrock. Consequently, observations regarding elements of dyke deposition allow us to establish the presence of different zonal blocks in granites along the strike. This is very important for subsequent stages of working the pit. These conditions make the otherwise laborious method of studying joints in the entire volume of rock massif considerably easier.

The analysis of elements of dyke deposition, mapped in the 'Selishe-2' pit with the help of profile micromagnetic survey and geological investigation methods, permitted us to conclude that in the granite massif being mined there exists three major joint systems of different ages:

1. System of vertical and steep (75°–95°) joints of submeridional (NW 350°, NE 5°) strike, confined by two sets of highly magnetic and narrow (up to 2 m) diabase dykes, opened up by the pit at a distance of 430 m from each other.
2. The system of more aged and flat (35°) joints of north-eastern (NE 15°–25°) strike, defined by the presence of weak magnetic but thick (up to 25 m) and flat dykes of metadiabases developed at a distance of 200 m from each other.
3. The system of inclined (45°) joints of sublatitudinal (NE 90°) strike, defined by several joints in granites of the north-eastern flank of the pit.

Using the data obtained, regionalisation of boundaries of granite pit 'Selishe-2' could be done accurately. In accordance with the suggested methodology, the parameters of effective fragmentation and seismically safe parameters of blasts were calculated for each zone in the pit horizons. The latter parameters ensured maximum energy saturation of the blocky massif during blasting due to the optimal interrelation of geometry and dimensions of borehole charge grids and their hook-up (delay intervals and sequence of blasting) within the boundaries of each zone. Such an approach to seismic microzoning of pits is the underlying principle in any typical blasting plan. The pit zoning is introduced in the typical plan as a separate division.

Seismic microzoning by means of a micromagnetic survey, if adopted, enhances the reliability of results in respect of pit zoning and makes the study of joints and blockiness of rock massifs easier. This is due to the usage of elements of dyke deposition which, in turn, leads to enhanced seismic safety and effective blasting operations.

1.2 Seismic Waves — A General Background (subsections 1.2–1.4
written in collaboration with S.N. Markelov)

The linear theory of elasticity is based on the assumption that deformation is a linear function of the applied stress (Hooke's law). In the nineteenth century, Poisson and Stokes proved that in a homogeneous isotropic elastic and unbounded space only an elastic longitudinal wave can travel. When vibrations occur in a confined volume of soil, or in the presence of a semi-space, two types of waves are generated: longitudinal compressional waves and transverse (shear waves). Rayleigh had shown that if the space is bounded by a plane, then along the free surface a third type of wave may be propagated. The faster of the two types of waves travelling in an elastic body is often termed the primary wave and the slower moving one secondary. They are designated as P (longitudinal) and S (transverse) waves respectively.

Dilation associated with P-waves does not include rotation and therefore a P-wave is further termed non-curling. As dilation includes compression (associated with volumetric modulus or the inverse quantity, compressibility), the P-wave is also known as a 'compressional wave'. As the particles move parallel to the direction of propagation of a P-wave, it is also termed longitudinal.

Rotation or distortion associated with S-waves includes no change in volume. Therefore, the S-wave is equivolumetric. As the rotation does not compress, but shears instead, the S-wave is also called a 'shear wave'. The particles connected with this wave move perpendicular to the direction of propagation. Therefore, the S-wave is also termed a transverse wave. With respect to the travel direction of particles, S-waves may be polarised as horizontal (horizontal movement of particles) or as vertical (vertical movement of particles). These are designated S_H and S_V respectively.

Surface waves travel slower than P- and S-waves and have been termed Rayleigh waves after the scientist who discovered them. They are indicated by the letter R. Other types of waves are also recorded on the surface of the earth: reflected, refracted, diffracted etc.

An analytical approach to the study of seismic vibrations should contain at least three of the following components: description of seismic sources; equations relating to motion, propagating in the medium after it has been generated at an arbitrary point; and the theory relating the description of seismic sources with partial solution established for equations of motion. Therefore, to uniquely define motion it is essential to know whether information about it in a particular part of the medium is observable in another part and whether it is adequate.

This problem has a specific solution for the elastic medium, i.e., conditions at the source (described by forces in all directions) and boundary conditions can be easily represented in a form that describes the uniqueness of the resultant motion.

Two methods are widely used to describe motions and the mechanics of motion in a continuous medium. The first is the Lagrangian approach, wherein a specific particle is considered whose initial position is given by its co-ordinates at a particular base time. The second is the Euler method which investigates motion of any random particles that pass by a given point in space. For most seismological problems, the linear theory of elasticity can be easily applied on the basis of the Lagrangian approach.

Let us introduce the concept of displacement and consider it as a function of time and space. Let us denote it as $a = a(x, t)$ so as to describe the vector distance of a particle at time t from point x, at which it was located at t_0. Since x does not vary with time, the velocity of particles is equal to $\dfrac{\partial a}{\partial t}$ and acceleration $\dfrac{\partial^2 a}{\partial t^2}$.

To determine the displacement $a(x,t)$, caused by a unidirectional point force applied with an amplitude depending on time, at the given point O in a homogeneous, unconfined, isotropic elastic medium, let us take O as the origin in the system of co-ordinates and axis x_1, as the direction of action of force. The equation to be solved to find the displacement is of the form

$$\rho a = \Phi + (\lambda + 2\mu)\nabla(\nabla \times a) - \mu\nabla \times (\nabla \times a), \qquad \ldots (1.1)$$

where ∇ is Hamilton's operator;

$\Phi = X_0(t)\,\delta(x)\,\delta_{ij}$ is volumetric force.

At initial conditions $a(x,0)$ and $x \neq 0$, the field of displacements, has components $a_n(x,t) = X_0 G$, where G is Greene's function of dynamic theory of elasticity; X_0 is the point force.

Solution to equation (1.1) for point force $X_0(t)$ in the direction x_j, is given by

$$a_i(x,t) = X_0 G_{ij} = \frac{1}{4\pi\rho}(3\gamma_i\gamma_j - \delta_{ij})\frac{1}{r^3}\int_{r/v_P}^{r/v_S} \tau X_0(t-\tau)d\tau + \frac{1}{4\pi\rho v_P^2} \times$$

$$\times \gamma_i\gamma_j \frac{1}{r}X_0(t - \frac{r}{v_P}) - \frac{1}{4\pi\rho v_S^2}(\gamma_i\gamma_j - \delta_{ij})\frac{1}{r}X_0(t - \frac{r}{v_S}) \ldots (1.2)$$

This formula (equivalent expression was obtained by Stokes in 1849) gives one of the important solutions to the problem of elastic wave investigation.

The relative value of the different terms in Greene's function depends on distance r between the source and receiver. Thus, the value $r^{-3} \int_{r/v_P}^{r/v_S} \tau X_0(t - \tau) d\tau$ varies as r^{-2} for sources with $X_0 \neq 0$ in an interval of time shorter compared to $r/v_S - r/v_P$ (for example, for an impulse source of even Greene's function). But the remaining terms in (1.2) are proportional to r^{-1} and as $r \to \infty$ they predominate over terms of r^{-2} order. Terms containing $r^{-1}X_0(t - r/v_P)$ and $r^{-1}X_0(t - r/v_S)$ are called displacements in the far-field zone. As the term with r^{-2} dominates over terms with r^{-1} as $r \to \infty$, the term containing $r^{-3} \int_{r/v_S}^{r/v_P} \tau X_0(t - \tau) d\tau$ is called displacement in the near-field zone.

The data used in geophysics are nearly always collected in a far-field zone, i.e., at distances where displacements in the far zone from (1.2) dominate. In certain cases, the seismic data used in engineering seismology are obtained from a near-field zone. However, it sometimes so happens that displacements in a far-field zone may be sufficiently large so as to cause seismic vibrational damage to buildings [2]. The P-wave in a far zone gives the displacement a_P in the following form from expression (1.2):

$$a_P^i(x, t) = \frac{1}{4\pi\Phi v_P^2} \gamma_i \gamma_j \frac{1}{r} X_0(t - r/v_P). \qquad \ldots (1.3)$$

This wave attenuates as r^{-1} along the given direction γ from the source. It has the form of oscillations, dependent on space-time combination $t - r/v_P$ and consequently, propagates with a velocity v_P, $(v_P^2 = (\lambda + 2\mu)/\rho)$. If $t_0 = 0$ and when $X_0(t)$ becomes for the first time non-zero, then r/v_P will be the time of entry of P-wave at r with displacement amplitude proportional to the applied force at a fixed time.

In the zone farther away, the wave happens to be longitudinal (sometimes, it is also called radial), i.e., particles in the wave move in the direction of propagation. The direction of displacement a_P at x is parallel to the direction γ from the source. This is arrived at from the property $a_P \gamma = 0$ obtained from expression (1.3).

The S-wave displacement a_S in the far-field zone is of the form

$$a_S^i(x, t) = \frac{1}{4\pi\rho v_S^2} (\delta_{ij} - \gamma_i \gamma_j) \frac{1}{r} X_0(t - r/v_S).$$

This wave along the given direction γ (γ is the unit vector, directed from source to receiver):
— attenuates at r^{-1};
— its time of arrival in $\bar{x}$ is equivalent to r/v_S and propagates with velocity v_S;

— has displacement amplitude, proportional to the applied force over time lag;
— has direction of displacements a_S in x, perpendicular to the direction of γ from the source.

Consequently, the S-wave in the far-field zone is a transverse wave, as the particles travel parallelly along the normal to the direction of propagation.

Displacement in the near-field zone is expressed, in accordance with (1.2) as,

$$a_N^i(x,t) = \frac{1}{4\pi\rho} \left(3\gamma_i\gamma_j - \delta_{ij}\right) \frac{1}{r^3} \int_{r/v_P}^{r/v_S} \tau X_0(t-\tau)d\tau. \qquad \dots (1.4)$$

In its formulation, the potential gradient of the P-wave as well as potential curl of the S-wave take part. In this regard, a_N is developed by both P- and S-waves. This displacement is neither non-rotational (i.e., having zero curl) nor solenoidal (i.e., having zero divergence). It follows that it is not always advisable to resolve elastic displacement into P- and S-components. Further, a_N carries both longitudinal and transverse displacements in the sense that the longitudinal component

$$a_N\gamma = \gamma_j \frac{1}{4\pi\rho r^3} \int_{r/v_P}^{r/v_S} \tau X_0(t-\tau)d\tau$$

and the transverse component

$$a_N\gamma' = -\gamma_j' \frac{1}{4\pi\rho r^3} \int_{r/v_P}^{r/v_S} \tau X_0(t-\tau)d\tau,$$

where γ_j is cosine of angle between direction of force and direction of a_P and a_P';

γ, γ' are longitudinal and transverse directions respectively.

Simple properties associated with displacements in a near-field zone cannot be applied for displacements in a far-field. However, the time of arrival and duration of displacement a_N at a fixed point of observation can be established.

Let $t = 0$ be the time $X_0(t)$ takes for the first time a non-zero quantity. Let us assume that $X_0(t)$ returns to zero for $t > T$. From the expression (1.4), it follows that a_N is the motion starting at point x at moment r/v_P (i.e., the time of entry of P-wave) and progressing up to $r/v_S + T$. It has the duration $(r/v_S - r/v_P) + T$. If $X_0(t)$ never returns to zero finally (i.e., T being non-finite), then displacement in the near-field zone takes place for an indefinitely long time (case of inelastic deformation).

Thus the main conclusions are: displacement in a far-field zone dampens as r^{-1} and is proportional to the velocity of particles at source. There are obvious distinct similarities in the displacements in near and far ones. The resultant statistical displacement induced in the medium by the shift in displacements, attains finally a specific value, is also damped as r^{-1}.

If the front of the incident wave (P or S) is not flat but curvilinear, then the condition of inversion to zero in a free surface is not satisfied when two different types of waves (P and S_V) are reflected. Instead of this, three types of waves form at the place of incidence:

$$P \rightarrow P + S_V + R \text{ and } S_V \rightarrow S_V + P + R.$$

The Rayleigh wave, thus, includes both dilation and distortion of the medium. This wave includes longitudinal movement of particles in the P-wave and transverse movement of particles in a vertical plane similar to the S_V-wave. For particles located on the surface, their combined motion makes an elliptical orbit in two-dimensional space on a distorted elliptical orbit in three-dimensional space. The amplitude of particles decreases faster with depth (equation of motion for the R-wave has an exponential function of depth), and the higher the frequency of the R-wave, the faster the attenuation of motion. Due to this, noticeable movement is observed near the surface. Hence R-waves are classified as surface waves, compared to P- and S-waves which are body waves. The velocity v_R of surface wave depends on the velocities v_P and v_S or on Poisson's coefficient, ν.

Nakano made a vital contribution to the theory of surface waves. He showed that the wavefronts of P and S, incident on a free surface, should be bent so that R-waves could be excited. One of the most interesting conclusions of Nakano is that the R-wave does not form near the epicentre and its origin is not precisely known; however, the approximate position of origin can be ascertained from the following equations:

for the source of P-waves

$$r_{PR} = v_R H (v_P^2 - v_R^2)^{-1/2}$$

for the source of S_V-waves

$$r_{SR} = v_R H (v_S^2 - v_R^2)^{-1/2},$$

where r is the epicentral distance;

H is the depth of source under free surface.

At these distances the R-wave is not created suddenly but grows gradually. At $r = r_{PR}$ or $r = r_{SR}$ the surface wave takes a shape characteristic to the R-wave and only at larger distances, where $r \gg H$, does this wave form entirely. Two more types of surface waves occur at a distance:

$$r_{PS} = r_{SP} = v_S H (v_P^2 - v_S^2)^{-1/2}$$

for sources of both P-waves and S_V-waves. If the source emits P-waves, then at a distance $r = r_{PS}$ the surface S_V-wave appears; if S_V-waves are emitted by the source, then the surface P-wave is developed. The geometry of these solutions is shown in Fig. 1. Nakano's locus diagrams are shown in Fig. 2. These diagrams indicate that, in contradistinction from the case of surface sources, the order of arrival of different waves does not necessarily include P, S and R.

As the theory of surface waves involves considerable mathematical complexity, it is not possible to represent their theoretical frequency characteristics. This difficulty has been the main reason for the non-publication until today of a complete analysis of the impact of R-waves in blasts. Researchers have concentrated only on the effect of a single or several corresponding physical parameters.

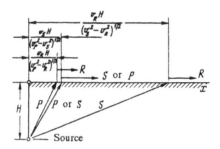

Fig. 1: Nakano's geometrical solutions for R-, P-, and S-waves and surface P-, S-waves, induced by a deep-seated linear source of P- and S-waves.

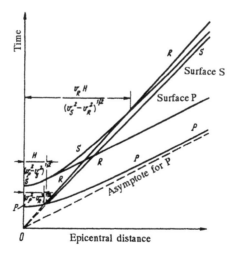

Fig. 2: Locus diagrams related to Nakano's solutions for R-, P- and S-waves and surface P- and S-waves.

Generally, the properties of a surface wave are as follows:

— Its propagation velocity v_R, satisfying the equation $R(1/v_R) = 0$, is lower than the velocity of transverse waves by several percents;

— It does not depend on the frequency; consequently, it is not subjected to dispersion;

— Wave attenuates with distance, as $r^{-1/2}$;

— Since the body waves are damped, as $\sim r^{-1}$ and the front ones as r^{-2}, the Rayleigh waves should predominate at considerably large distances;

— Phase lag in the R-wave is equal to $\omega r/v_R$ and it does not depend on depth H since the locus diagram of the wave is a straight line;

— Amplitude of R-wave is an exponentially decreasing function of H and ω;

— Particles in R-wave travel in elliptical trajectories; on the surface their movement is regressive;

— Below a specific depth, depending on frequency, the S_V component is predominant in particle motion in the Rayleigh wave and the travel path becomes elliptical.

General concepts regarding the theory of wave processes in elastic media have been given here. But in real conditions the patterns of formation and propagation of elastic waves are highly complicated by many and varied conditions both at the source of elastic vibrations and in their travel paths. This is particularly applicable to seismic waves arising from blasts in open-cast and underground mines. Though vibrations in such a case follow the main laws of the theory of elastic waves, their analytical description is very difficult and often not possible due to a whole host of geo-mining and technological factors influencing the wave processes.

1.3 Impact of Natural and Technological Factors on Seismic Effects of a Blast

Blasting conditions: In industrial blasts the wave picture is extremely complex. This is due to the prevailing geo-mining conditions on the travel path of blast-induced seismic vibrations and also due to the special nature of the blast itself as a source of elastic waves.

In describing such a source we can only consider approximations of the models as applied to the properties of the medium in which blasting takes place. In actual conditions, various endogenous factors (type of explosives, weight, construction and shape of individual parts in a charge and the total charge in the block being blasted, initiation scheme etc.) as well as external factors (properties of rocks, availability of free face, line of least resistance, depth of charges) etc., directly or indirectly influence the blast.

Many papers have been published in the USSR and abroad which consider the effect of these factors on the intensity of the source of seismoblast vibrations and the entire seismic effect of blasts. Most of the investigations are site-specific, i.e., their results are only applicable to the mines in which the study of the seismic effect of blasts was conducted. At the same time, generalisation of the data is of paramount importance for practical purposes as it facilitates the selection of principal directions in studying the seismic effects of blasts at specific locations and in forecasting the influence of any of these factors in industrial blasting operations.

Construction of explosive charge: The properties of the explosive used in blasts primarily influence the intensity of the source of seismic vibrations. The effect of explosive properties is expressed in the form of distribution of energy expended for fragmentation, for exciting seismic vibrations, in the changes of duration of blast effects on the surrounding medium and frequency composition of vibrations. Explosives having low velocity of detonation (VOD) are preferred for conducting blasts to produce reduced seismic effects. Less seismically active explosives are recommended for use: igdanites, ifzanites, granulites and grammonites. For example, grammonites and granulites are 20–25% less seismically active than ammonite 6-ZhV, while for trotyl and allumotol the activity is 20–25% more. Explosives with higher VOD generate significant vibrations. In their spectra, higher frequencies predominate, which absorb a major part of the energy. Therefore, while selecting explosives due consideration should be given to the requirements of fragmentation and absorbing properties of the surrounding rocks at different phases in the frequency spectra of oscillations.

The influence of explosive properties on the seismic effects of a blast can be reduced by using charges of variable density along the drill hole length and also by the use of charges with inert gaps. By so doing we can lessen the seismic effects of a blast to safe limits.

The most effective method of reducing blast-induced seismic effects as well as enhancing the quality of fragmentation is to use inert and air gaps and also inert stemming. It has been established that the intensity of vibrations is reduced by 1.2–2 times, depending on the properties of surrounding rocks, when charges with air gaps are used. However, the use of such charges reduces the seismic effects only at specific ratios of volume of air gaps to the entire charge volume in a particular deposit. This ratio is about 0.3–0.4.

Increasing the height of air gap in subdrilling while maintaining charge weight results in considerably reduced intensity of longitudinal wave, gives a better shape to slope and reduces the extent of damage to the floor level bench. Interhole delays enhance the effectiveness of blasting explosive charges having inert gaps.

The diameter of drill holes undoubtedly influences the seismic effects of a blast. According to researches carried out on this aspect, the seismic effects of a blast are reduced when hole diameters are reduced. The relationships characterising the change in seismic effects of blasts with variation in drill hole diameters have been analysed [23].

Compared to hole diameter, the length of stemming exerts a lesser influence on the seismic effects of blasts. Stemming only indirectly affects them by redistributing the energy and directing it more towards fragmentation. The change in quality and quantity of stemming has a bearing on the seismic effects, particularly in monoliths and weakly fissured rocks or soils. This is due to the increased effect of products of expansion in a blast besides the effect of compressive blast waves in the massif. The quantity of stemming is governed by the requirements of fragmentation and reduction of fly rock. It should be not more than half the length of a charge column in a drill hole and not less than twenty times the hole diameter [29].

Conditions of placing charge: The conditions of charge placement considerably influence the seismic effects of a blast. Maximum seismic effects are observed in blasts conducted in a confined medium. The depth of charge placement plays a vital role since with any increase in depth, the intensity of vibrations also increases. As the number of free faces increases, the vibration velocity of rocks decreases. In such a case seismic effects may be reduced by as much as 4–5 times compared to blasting in a confined medium.

In a series of investigations, the change in seismic effects of a blast due to change in bench height or length of hole charge was considered. It was established that relatively rapid growth of particle velocity is noticed when the bench height is increased from 10 to 20 m. Further increase in height results in reduced acceleration of displacement velocity and finally to total stoppage. When the bench height is increased by 3.5–4 times, the displacement velocity increases by only 30%. The enhanced intensity of seismic vibrations can be explained by the increased consumption of explosives per unit time of blast and also by the lengthening of charge. When the seismic effects are determined by the weight of the charge per unit length of hole, the increase in charge length considerably influences the transition to cylindrical symmetry. Slowing down and stoppage in augmentation of displacement velocity are observed as the bench height is increased, when such a level of relative lengthening of charge is achieved.

Properties of rocks: The seismic effects of any blast depend to a large extent on the properties of rocks at the site of blasting, besides the constructional details of charges and the technological conditions of their placement.

The seismic effects of any blast are closely related to breakage, fragmentation and displacement of the surrounding rocks. That part of the energy spent on generating elastic vibrations is determined based on physicomechanical properties of rocks and varies over a wide range, from 0.01 and even 0.001% to 1% of the total blast energy.

Another important property is the acoustic rigidity of rock. Placing a charge in a medium of lower acoustic rigidity reduces the seismic effects of a blast. A blast in rocks of relatively greater acoustic rigidity produces 3 times more seismic energy at the source boundary, compared to blasts in rocks with lower acoustic rigidity.

Blasts in clays, marlstones and salts cause maximum ground movement due to the seismic wave. While blasting in hard rocks, the extent of development and nature of fissuring affect the seismicity considerably. As the specific fissuring index increases, the seismic effect in large blasts reduces. At the same time, a vital role is played not only by the number of fissures, but also by the extent of their opening, filling by secondary products and spatial orientation. In certain cases, maximum displacement velocity is noticed in blasts in rocks with minimum sizes of rock units. The spatial disposition of fissures also greatly influences the seismic effects of a blast. By properly orienting the drill hole grid, the fragmentation and intensity of elastic vibrations can be regulated.

Change in the physicomechanical properties of rocks at the site of blasting also influences the frequency composition of blast-induced vibrations. In rocks with a low value of acoustic rigidity, lower frequencies dominate compared to rocks with higher acoustic indexes.

Due attention needs to be paid to the impact of a prestressed state of rocks on the seismic effects of a blast.

Short Delay Blasting (SDB): The use of SDB in shot-holes and drill holes has been extensive since the 1960s. It has advantages over instantaneous blasting —possible control of fragmentation, fly rock and trajectory of throw of the blasted rock mass.

From the point of view of reduced seismic damage, SDB is a very effective method and in certain cases reduces the seismic effects of blasts to that of the effect of the explosive weight used for a single delay. This makes it possible to undertake large-scale industrial blasts according to the technological demands of mining enterprises.

Use of SDB in confined conditions of industrial and civil structures is particularly important. In such conditions different technological measures are adopted which facilitate conduction of blasting operations without damaging structures in close proximity. Such measures include orienting the front of seismoblast vibrations, controlling seismic effects by sequential blasting of different delays etc.

Similar to the parameters of instantaneous blasting, the SDB parameters are selected according to the properties of the rocks at the blasting site, technological schemes of initiating the blast, location of industrial and residential structures relative to the site of the blast etc.

Generally, if an optimal delay time is selected (without taking into account the fragmentation factor), based only on the requirements to reduce interference in the propagation of seismoblast waves in a zone having structures to be protected, and if the 'undercutting' of neighbouring charges in a block is ruled out, then the optimal delay time varies within fairly wide limits, from 10 to 80 ms and more. It is necessary to make the delay time compatible with the type of rocks, their acoustic properties and period of oscillation of waves. For example, in the breakage of granites by blasting, the optimal delay time happens to be 15–35 ms. As the acoustic rigidity of rock decreases, the delay time is increased. In limestones, the optimal delay time is 20–50 ms. In weak rocks, delays with large intervals (50–80 ms) are preferred. In very hard rocks (some types of limestone), the scattering of delay time intensifies the seismic effects while in weak rocks the opposite is seen. To avoid the interference of seismic waves, it is necessary to see that the delay interval exceeds the duration of the positive phase of the seismic wave [24].

The use of SDB allows the introduction into blasting practice of different schemes of initiation — row-wise, wedge type, radial, wavy etc.

Their effect of the seismicity of a blast is caused by the general redistribution of blast energy on generating elastic vibrations and breaking and displacing the rock. This effect is related to changes in line of least resistance in the block, angle of free action of charges, availability and formation of free faces in the blast. The diagonal row and trapezoidal cut schemes are widely used schemes of initiation from the point of view of maximum reduction in the seismic effects of a blast. It should be noted that the use of various SDB schemes with optimally selected parameters allows the reduction of seismic effects of the entire blast to that of a single delay blast effect.

The seismic effects of a blast can be controlled even to a much greater extent by changing the commutation of charges so as to change the direction of wavefront. In almost all the literature that analyses the spatial disposition of charges in a block, a lower degree of vibrations is recorded for the butt end of the block being blasted or in a series of drill holes compared to vibrations in the length-wise portion of the block. The direction of the blast-induced wavefront in such cases depends on the direction of initiation of the chain of charges, distance between them and the velocity of longitudinal waves in the massif. The influence of direction of propagation of blast-induced vibrations is detailed in the chapter concerned with the propagation of seismoblast waves. Here it is to be mentioned that after firing an elongated dispersed charge the velocity of rock displacement at points along the line of drill holes is constrained at the lower bound by the velocity of single concentrated charge and at the upper bound by the velocity of the entire concentrated charge in a direction, normal to the row of drill holes in case of instantaneous blasting. Hence any structure to be protected may be located at the flank of the block to be blasted because it will direct the detonation from the structure side and thereby reduce the velocity of ground vibrations by 2–3 times in its neighbourhood.

The trend towards larger blasts using a large number of delay intervals in a number of blocks, leads to the overall lengthening of blast cycle duration and consequently to the enhancement of seismic effects on structures to be protected.

To reduce the seismic effects of a blast, the total time of blast duration should be reduced by optimally selecting delay intervals and the number of delay groups.

In some publications, recommendations have been given to create a meeting front of seismic waves with out-of-phase vibrations and other measures.

1.4 Special Features in the Propagation of Blast-induced Seismic Waves

Blast-induced waves usually follow the general laws governing the generation and propagation of seismic waves in the earth's crust. Longitudinal, transverse and surface waves are also noticed during industrial blasts. Sometimes sufficiently intense waves are noticed, which are created by reflection, refraction and diffraction of oscillations.

However, the wave picture of seismic vibrations induced by blasting operations has its own peculiarities. Firstly, the blast as a source of seismic vibrations has a highly complicated structure, not easily amenable to mathematical description. This is mainly due to the use of complicated initiation schemes aimed at obtaining the required fragmentation, short delay blasting, spatial concentration of charge within the block as well as in other benches and sections of mines. Due to this, the waves generated by various charges — with different delays, from different blocks and levels — overlap. In the process, the arrival sequence of various types of waves is often disturbed, which renders the interpretation of seismograms as well as the assessment of kinematic and dynamic characteristics of waves more difficult. Secondly, geo-mining conditions are extremely diverse and vary all along the propagation path of seismoblast vibrations right from the blast site to the area of concern. One should take into account the fact that buildings are often situated in a zone in which the seismoblast waves are yet to develop fully. The difficulty arises therefore in assessing the attenuation parameters of waves and in forecasting their intensity near residential and industrial structures.

Characteristics of seismoblast waves: Among the elastic waves of high intensity, longitudinal, transverse and surface waves are distinguishable. Their parameters vary depending on charge mass, distance to the blast site, orientation of wavefront relative to the direction towards the observation point, conditions prevailing in the travel path of waves etc.

Nearer to the blast site, body waves (longitudinal and transverse) predominate. They are characterised by relatively high frequencies (of the order 10–40 Hz) and rapid attenuation compared to the surface waves. The surface wave follows next (in the time sense). It is characterised by weak damping over distance, large amplitudes and lower (of the order 2–8 Hz) frequencies. The typical feature of a surface wave originating open-pit blasts is that it develops not only due to the interaction of longitudinal and transverse waves at a distance determined mainly by the charge depth (as in the case of a deep-seated source in seismology and seismic exploration), but also due to the dome-shaped heaving of the massif's surface in the immediate vicinity of the blast site [26].

Other types of waves are also distinguished, whose sources are the boundaries of jointed zones, zones of elastic deformation. But these waves possess localised features and play a significant role only in a particular zone with typical properties of rocks and constructional peculiarities of the source of blast vibrations.

In the case of complex geological structure of a massif having layers and bodies with enhanced acoustical rigidity and propagation velocity of waves, reflected and refracted waves could originate as determined by the geological conditions at depth and on the surface. These waves have intensities comparable to that of longitudinal and transverse waves.

The dynamic features of blast-induced seismic waves largely depend on the changes in distance from the blast site. In the nearer zone a sharp increase in the displacement velocity occurs initially, with subsequent rapid attenuation. At greater distances, in the elastic zone, vibrations have a shape similar to that of a sinusoidal wave.

The grouping of zones according to deformations of a rock massif is a widely subscribed concept among investigators: near zone — zone of irreversible deformations; middle zone — zone of elastoplastic deformations; far zone — zone of elastic deformations. In the near zone, seismic vibrations attenuate due to irreversible deformations. The nature of dampening of waves varies as they move away from the blast site and at distances equivalent to 2–3 times the wavelength, elastic deformations are observed. In terms of forecasting the seismic effect, the grouping of zones based on the type of waves (body waves, surface waves) is preferable (without considering the zone of inelastic deformations). These zones have typical characteristics corresponding to the wave types, their attenuation, frequency composition and intensities.

Effect of rock properties: The energy of seismoblast vibrations attenuates depending on the structure of rocks and soil. Usually, the displacement velocity depends on the acoustic rigidity of rocks at various scaled distances. Maximum attenuation of displacement velocity occurs in weaker and fractured rocks. Clayey and watery soils possess higher seismicity.

The fissuring nature of hard rocks greatly determines the propagation of seismoblast vibrations. The attenuation coefficient depends on the density of joints as well as on the degree of their opening and properties of filling material.

Seismic anisotropy of the massif plays a major role in the propagation and dampening of seismoblast waves in jointed massifs. This anisotropy can be explained by the varying influence of jointing systems on attenuation of the high frequency part in the spectrum of oscillations. Therefore, to forecast the intensity of seismic vibrations, it is necessary to take into account jointing in different directions. However, the dampening of seismoblast waves in a massif of locally jointed rocks is complicated and hence the seismic effect cannot be quantified according to known formulae.

Vibrations propagate readily in the direction parallel to bedding planes. In this case the intensity of oscillations in the wave, based on the massif structure and distance, is 1.4–3 times more compared to the intensity of oscillations during wave propagation across the bedding planes.

Structure of overlying strata: Loose formations on a hard rock foundation intensity the seismic effects of a blast, as they happen to be rock layers with reduced acoustic rigidity and have high velocity of displacement and larger oscillation periods compared to the massif. At the same time, a rock layer with higher rigidity and located on weaker rocks happens to be a good conductor of waves for seismic vibrations.

Influence of depth of blasting: Mutual location of surface relief: The amplitude of seismic oscillations, distinct to the source, depends on the blast energy and the source depth. Increase in the power and depth of blast results in a complicated picture of seismic oscillations and the appearance of secondary (reflected) waves with a longer duration of oscillations.

The distance at which the surface waves are formed is closely related to the increase in depth of blast. The surface waves carry a major part of the seismic energy of the blast. The radius of formation zone of a surface wave is 3–5 times the depth of blast and is determined by the ratio of kinematic parameters of longitudinal and transverse waves in actual conditions of propagation of vibrations during blasts.

There is a relationship between coefficients of proportionality in the formulae relating displacement velocity to the scaled distance and blast depth. This is due to variations in geo-mining conditions with increasing depth of a pit [9].

The surface relief between blast site and observation point largely influences the seismic effect of a blast. Along the slope of a pit, in most cases, the magnitude of vibrations is 1.5–2 times more than the vibrations observed in the horizontal area at a given distance. The intensity of vibrations along the slope surface is also not uniform. Near the bench edge more intensive vibrations are noticed compared to the opposite part of the flank and, in the mid-portion, the intensity of vibrations changes only marginally [27].

This problem has been elaborated in subsequent chapters. Here attention is drawn to the influence of a pit flank on the interaction of body waves and formation of surface waves behind the bench.

Presence of screening zones: The existence of a mined-out area or disturbed zones between the blast site and the protected buildings influences significantly the formation and propagation of seismoblast waves. A mined-out area (excavated part of the pit, working trench etc.) has a screening effect on seismoblast waves. The intensity of vibrations reduces by 1.5–2 and in some cases even 2–3 times compared to the travel of waves through an undisturbed massif. In order to reduce seismic effects of a blast, this phenomenon can be utilised in planning blasting operations, by ensuring that a mined-out area is made available between the blast site and protected structures. This can even be undertaken by creating artificial barricades in the travel path of seismoblast waves.

Methods for calculating seismic effects of a blast: The principal criterion for assessing the seismic effects of a blast in a rock massif is the velocity of vibrations (displacement velocity of soil particles), which depends more on the conditions of wave propagation than on the amplitude of oscillations and acceleration. It is directly proportional to breakage due to the blast and determines the energy of seismic waves. It allows us to comprehensively implement the technology of blasting operations in industrial conditions.

Currently in the USSR and abroad, seismic methods are used to quantitatively assess the seismic effects of a blast. Among these the prominent one is the solution of relationship of velocity (in the medium due to blasting of an explosive charge) as a function of distance from blast site to observation point and explosive charge weight $u = kC^n r^\beta$, where k, n, β are empirical coefficients, dependent on conditions of blasting and propagation of seismoblast vibrations.

Such relationships take into account directly the entire set of blasting parameters influencing the seismic effects of a blast. Such parameters are constructional features of an explosive charge, SDB parameters, initiation schemes, rock properties and geo-mining conditions. In such an exercise, the measured values may deviate much from approximated ones. This can be explained by the inevitable and uncontrollable variation of many blasting parameters.

Academician M.A. Sadovski's method is widely used in the USSR. It is based on the similarity criteria that establish a relationship between displacement velocity and scaled distance or scaled weight of charge

$$u = k(\overline{r})^n \text{ or } u = k(\overline{c})^n, \qquad \ldots (1.5)$$

where k is the proportionality coefficient;

n is the index of damping level;

$\overline{r} = r/\sqrt[3]{C}$ is the scaled distance;

$\overline{C} = \sqrt[3]{C}/r$ is the scaled weight of the charge;

r is the distance between the blast site and the observation point, m;

C is the weight of charge, kg.

It should be noted that even for a homogeneous deposit the propagation of seismic waves is quite distinct at different times, due to the extremely varying set of geo-mining and technological conditions.

In most cases the results obtained by different researchers are justified only for conditions in which the experiments were conducted. These experiments are significant in highlighting the directions for further research. They enable us to select and assess the more important factors that influence the seismic effects of a blast in the prevailing conditions of a mine as well as to find more effective means of reducing such effects.

Permissible displacement velocities: The permissible velocity of oscillations in the most intensive wave is the generally accepted criterion in the USSR for assessing the seismic effects of blasts. Permissible velocities at foundations of structures and buildings are determined by their constructional features, present state and dynamic characteristics.

Industrial and residential buildings and structures are grouped into classes. Depending on the class to which these protected buildings belong and their present state, the permissible velocity of oscillations for the given object is established.

Seismic effect induces breakage of bonds between individual constructional elements, thus causing often damage to buildings and structures. Here, the evaluation of deformation properties of building materials is quite significant. For example, as per V.L. Lavrinenko, the permissible velocity of vibrations in concrete is about 40–120 cm/s. For reinforced concrete under one-time loading, the critical velocity of vibrations is 130 cm/s, under multiple loading 50 cm/s, and with large axial tension 20 cm/s. The joints between individual elements in a building are damaged even at lower values of displacement velocity.

The range of permissible displacement velocities is quite wide and based on constructional features, state and purpose of the building may vary from 0.5 to 10 cm/s and even more. A displacement velocity up to 5 cm/s is considered a relatively safe zone, as at higher velocities cracks develop, plaster peels off, old cracks open up etc.

While determining permissible velocities of vibration, it is essential to take into account their frequency spectrum, as waves of different frequencies having similar velocities of displacement cause damage of different magnitudes. High-rise buildings move as a whole at frequencies of 50 Hz. At 10 Hz and more, bending and tensile stresses dominate in the building elements while at 5–10 Hz both types of loading occur. Long buildings split into separate blocks usually via cracks. This is due to the action of bending forces as different parts of the building oscillate at different times. Thus, in selecting the permissible velocity it is essential to take into account the frequency of vibrations or to establish various velocities for different frequency ranges.

To solve certain practical problems, it is necessary to also introduce the maximum velocity of vibrations corresponding to the boundary limits for safeguarding buildings or structures. If the velocities of vibrations in the rock massif are higher than permissible, then the protection of buildings becomes probabilistic in nature. In such a case, the extent of protection can be assessed by the constructional features of the building. The critical velocity of vibrations can then be defined by such ultimate permissible velocity exceeding which the probability of protecting buildings and structures is less than 0.5. The ultimate permissible velocity can be assumed as two times higher than the permissible value. Blasting operations can be undertaken only in particular cases where the vibrations are expected to touch permissible values of vibration velocities. For example, in emergency conditions one-time blasting operation is permitted even at the ultimate velocity of vibrations. In each case, the situation is assessed from the economic viewpoint of restoring the buildings and structures [24].

Establishing the permissible velocities of vibrations that ensure the safety of buildings is of paramount importance in the selection of quantitative blast parameters, permissible explosive charge weight and safe distances in industrial situations.

2

Frequency Spectra of Seismic Vibrations

2.1 Spectral Characteristics of Blast-induced Seismic Waves

The problem of studying seismic vibrations can be resolved to a large extent by using spectral analysis of seismic waves which enables supplementation and refinement of the concepts concerning the mechanism of these vibrations. Such analysis further facilitates assessing the behaviour of buildings and structures when subjected to loading. The seismicity in a given region is specified by the predominant periods and spectral composition of vibrations in rocks, their dependence on the intensity of blast effects, epicentral distance, depth of placing charge, soil and geological conditions and nature of attenuation of seismic waves.

The practical significance of studying the spectral composition of seismic vibrations lies in the fact that, knowing the predominant periods of vibrations, it is possible to select periods (frequencies) of natural oscillations of buildings and structures so as to avoid significant build-up of amplitudes of vibration during blasts. This aspect is particularly useful in designing structures, installations and other constructions in which the waves dampen very little.

The maximum amplitude of blast waves and duration of their action are generally used in the calculations of dynamic loads on structures during blasting operations. By varying these parameters, using an appropriate method, the blast effect on the rock massif is controlled. However, knowledge of these two parameters of wave is not adequate.

The laws of variation of spectral characteristics depending on certain influencing factors are given below.

At a distance r from the blast source, the stress wave is described by the analytical function

$$\sigma(t) = f(t). \qquad \ldots (2.1)$$

Let us introduce the spectral density of blasting effect

$$S(j\omega) = \int\limits_{-\infty}^{\infty} f(t)\, e^{-j\omega t} dt. \qquad \ldots (2.2)$$

As is known from spectral theory, there exists a relationship between (2.1) and (2.2):

$$\sigma(t) = \frac{1}{2\pi} \int\limits_{-\infty}^{\infty} S(j\omega)\, e^{j\omega t} d\omega. \qquad \ldots (2.3)$$

Formulae (2.2) and (2.3) are very important in spectral theory. They are a pair of Fourier transforms, linking the function of real time $f(t)$ and the complex function of frequency $S(j\omega)$.

Formula (2.3) is a Fourier integral in a complex form. Its essence is that the function $f(t)$ is represented as a sum of sinusoidal components. Therefore, it can only be represented as the sum of an infinitely large number of infinitely smaller vibrations having infinitely closer frequencies. The complex amplitude of each vibration is infinitely small and is equal to

$$dC = \frac{1}{\pi} S(j\omega) d\omega.$$

Formula (2.3) can be written in a real form. Then only the positive frequencies are integrated. Denoting $S(j\omega) = A(\omega) + jB(\omega)$,
we obtain

$$\sigma(t) = \frac{1}{\pi} \int\limits_{0}^{\infty} \Big[A(\omega)\cos\omega t - \dot{B}(\omega)\sin\omega t \Big]\, d\omega, \qquad \ldots (2.4)$$

where $A(\omega)$ is the even function
and $B(\omega)$ is the odd function.

Formula (2.4) can even be written as

$$\sigma(t) = \frac{1}{2\pi} \int\limits_{0}^{\infty} \Big[S(j\omega)\, e^{j\omega t} + S(-j\omega)\, e^{-j\omega t} \Big]\, d\omega.$$

The sum of conjugate values, equivalent to the doubled real part, is given in brackets. Therefore,

$$\sigma(t) = \frac{1}{\pi} \int\limits_{0}^{\infty} S(j\omega)\, e^{j\omega t} d\omega.$$

Let us introduce one more refinement. The function under integral in (2.3) expresses a separate infinitely small component, i.e., vibration with infinitely small amplitude dC:

$$\frac{1}{\pi} S(j\omega)\, e^{j\omega t} = dC\, e^{j\omega t}.$$

From this we find that

$$S(j\omega) = \pi \frac{dC}{d\omega}.$$

Thus $S(j\omega)$ does not directly denote amplitude but rather the so-called spectral density. However, this aspect is ignored and $S(j\omega)$ is termed as a complex spectrum of non-periodic function and the absolute magnitude (modulus) of this quantity $|S(j\omega)| = S(\omega)$ is simply termed the spectrum.

The spectral characteristic of a blast pulse, introduced in formula (2.2), enables assessing the variation in its parameters at different distances from the centre of blast, taking into account the dissipational losses.

Apart from the spectral function, the following characteristics reflect the blasting process in depth:

(a) Spectral energy

$$E_S = \frac{1}{\pi} \int\limits_0^\infty S^2(j\omega)d\omega. \qquad \ldots (2.5)$$

(b) Operational duration of pulse $\Delta\tau$, which signifies the interval of time where the major portion of pulse energy is concentrated. This is derived from the equation

$$\int\limits_{t_0-\Delta\tau/2}^{t_0+\Delta\tau/2} \sigma^2(t)dt = \mu_p \int\limits_{-\infty}^\infty \sigma^2(t)dt = \mu_p A_t, \qquad \ldots (2.6)$$

where μ_p is the proportion of total pulse energy expended in the time interval $\Delta\tau$ (usually $\mu_p = 0.9$);

$A_t = \int\limits_{-\infty}^\infty \sigma^2(t)dt$ is the quantity proportional to the total pulse energy;

(c) Operating width of spectrum Δf is found out from

$$\int\limits_0^{2\pi\Delta f} S^2(j\omega)d\omega = \mu_p \int\limits_0^\infty S^2(j\omega)d\omega; \qquad \ldots (2.7)$$

(d) Coefficient of blast pulse shape $\eta_p = \Delta f \Delta\tau$.

This characterises not only the pulse shape, but also its distortion while passing through the measuring track within the limited operating frequency band;

(e) Frequencies of significant harmonic components, determined from the spectral curve $S(\omega)$.

Let us consider the influence of blast wave parameters on the amplitude-frequency spectrum in the light of the above spectral characteristics.

Blast waves act on buildings in the same way as transient dynamic loads. While performing calculations on dynamic response of buildings, the actual laws of load variations in time are replaced by simplified calculations. The law of variation of dynamic load of stress waves over time is often used in the form

$$\sigma(t) = \begin{cases} 0, & t < -\tau_1; \\ \sigma_{max}(1 + t/\tau_1), & -\tau_1 \leq t \leq 0; \\ \sigma_{max}(1 - t/\tau_2), & 0 \leq t \leq \tau_2; \\ 0, & t > \tau_2; \end{cases} \qquad \ldots (2.8)$$

where σ_{max} is the maximum value of stress waves;

τ_1, τ_2 is the time of build-up and reduction of stresses, correspondingly.

As per formula (2.2), the spectral density of blasting effect is written in the form

$$S(j\omega) = \frac{\sigma_{max}}{\omega^2} \left(\frac{1 - e^{j\omega\tau_1}}{\tau_1} + \frac{1 - e^{-j\omega\tau_2}}{\tau_2} \right).$$

For convenience of study of the relationship between frequency spectrum and stress-wave parameters, let us introduce the following notations:

$$\tau_1 = a_i\tau; \qquad \tau_2 = b_i\tau,$$

where τ is the total duration of the blast effect, s.

Using the above notations, the spectrum of dynamic load represented in formula (2.8) would be of the form

$$S(\omega) = |S(j\omega)| = \frac{\sigma_{max}}{a_i b_i \tau \omega^2} [2(a_i^2 + b_i^2) + 2a_i b_i + 2a_i b_i \cos(a_i + b_i)\omega\tau -$$

$$-2(a_i + b_i)(a_i \cos b_i \omega\tau + b_i \cos a_i \omega\tau]^{1/2}. \qquad \ldots (2.9)$$

After simple manipulations, (2.9) becomes

$$S(\omega) = \frac{\sigma_{max}}{a_i b_i \tau \omega^2} [1 + a_i^2 + b_i^2 + 2a_i b_i \cos \omega\tau -$$

$$-2(a_i \cos b_i \omega\tau + b_i \cos a_i \omega\tau)]^{1/2}. \qquad \ldots (2.10)$$

For a particular case when $a_i = b_i = 0.5$, from expression (2.10) we get

$$S(\omega) = \frac{4\sigma_{max}}{\tau \omega^2} \left(1 - \cos \frac{\omega\tau}{2} \right).$$

The analysis in (2.10) shows that the spectrum depends on the maximum value of blast loading σ_{max} and its time τ of action. Further, this time influences the frequency composition of the spectrum.

The frequency spectrum curves, at constant duration of action $\tau = 10^{-2}$ s and various values of a_i determining the time of wave build-up, are shown in Fig. 3. The values of $S(\omega)/F_p$, where $F_p = \sigma_{max}\tau/2$ area of stress diagram, are

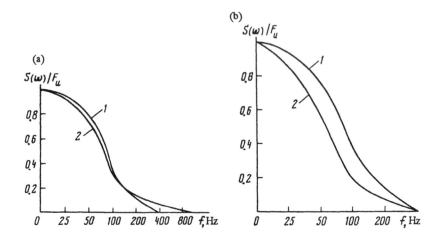

Fig. 3: Frequency spectral curves of blasting effects:

(a) with different times of wave build-up; (b) with equal time of wave build-up.

(a): $1 - \tau_1 = 5 \cdot 10^{-4}$ s; $\tau_2 = 95 \cdot 10^{-4}$ s; $a_i = 0.05$; $b_i = 0.95$;
$\quad 2 - \tau_1 = 25 \cdot 10^{-4}$ s; $\tau_2 = 75 \cdot 10^{-4}$ s; $a_i = 0.25$; $b_i = 0.75$;

(b): $\tau_1 = 25 \cdot 10^{-4}$ s; $\tau_2 = 75 \cdot 10^{-4}$ s; $a_i = 0.25$; $b_i = 0.75$;
$\quad 2 - \tau_1 = 25 \cdot 10^{-4}$ s; $\tau_2 = 125 \cdot 10^{-4}$ s; $a_i = 0.17$; $b_i = 0.83$.

plotted on axis y while the values $[f = \omega/(2\pi)]$ frequency, Hz are plotted on axis x

The value of $S(\omega)$ at $\omega = 0$ can be found out by the formula

$$S(0) = \lim_{\omega \to 0} S(\omega) = \sigma_{max} \tau / 2.$$

Thus at $\omega = 0$, the amplitude of spectrum is numerically equal to the area of dynamic loading diagram and does not depend on the ratio between parameter a_i and b_i.

From Fig. 3(a), it follows that with reduction in the parameter a_i (or length of leading front of stress wave), the spectrum moves into the region of higher frequencies. Thus at $a_i = 0.05$, $\tau_1 = 5 \cdot 10^{-4}$ s (curve 1) frequencies up to 1600 Hz are present in the spectrum, while at $a_i = 0.25$, $\tau_1 = 25 \cdot 10^{-4}$ s the amplitude of spectral density is equal to zero even at a frequency of 400 Hz.

The physical sense of parameter a_i lies in the fact that it indirectly establishes the rate of application of blasting effect (rate of loading). Actually, this velocity in the given case, can be found out from the formula

$$\frac{d\sigma}{dt} = \frac{\sigma_{max}}{\tau_1} = \frac{\sigma_{max}}{a_i \tau}.$$

The frequency spectral curves of the blasting effect having constant duration of the leading front (similar rate of loading) and different time of reduction in stresses τ_2, are shown in Fig. 3,b.

Variation in the time of fall of stresses τ_2 does not influence the frequency composition of the spectrum. Consequently, the frequency composition in the case of a blast effect is determined only by the duration of the leading front τ_1.

2.2 Amplitude-Frequency Spectra for Different Regimes of Interaction of Stress Waves

The study of spectral characteristics during the interaction of stress waves from several sources of blast is quite interesting. Such a study aims at selecting the optimal schemes of blasting operations. For this purpose let us consider a set of schemes in which the blast waves interact and let us evaluate their effectiveness with the help of spectral characteristics.

Let us study the blast effect of two explosives charges placed at a specified distance from each other. The law of variation in blast loading due to each charge is assumed to be a right-angled pulse.

Let us consider the first scheme, for which as a result of superposing wavefields at a distance r from the blast sources, a sum total pulse is obtained in the manner shown in Fig. 4a.

The spectrum of this pulse, as per (2.2) is described by the formula

$$S_1(\omega) = 2\,\sigma_{max}\tau\frac{\sin\omega\tau}{\omega\tau},$$

where σ_{max} is the maximum amplitude of the pulse;

τ is the duration of pulse;

$\omega = 2\pi f$ is the angular frequency.

The second scheme of interaction of blast pulses is shown in Fig. 4,b. The spectrum of resultant pulse is written in the form

$$S_2(\omega) = 2\sigma_{max}\tau\frac{\sin\omega\tau/2}{\omega\tau/2}.$$

Let us establish the duration of acting pulse $\Delta\tau$ and operating width of spectrum Δf as per formulae (2.6) and (2.7). For the first scheme, $\Delta\tau_1 = 1.8\tau$, $\Delta f_1 = 0.9/(4\tau)$; for the second scheme, $\Delta\tau_2 = 0.9\tau$, $\Delta f_2 = 0.9/2\tau$; the parameter η_p is the same in both cases, $\eta_p = 0.405$.

For the first scheme, $\Delta\tau_1 = 2\Delta\tau_2$ and $\Delta f_1 = 0.5\Delta f_2$, i.e., the acting duration of pulse is twice more than that in the second scheme, and the spectral width is less by twice as much. As the lower frequency spectrum undergoes less distortion than the spectrum saturated with higher frequency harmonics, adoption of the first scheme is preferred for achieving a better compaction.

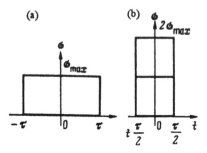

Fig. 4: Schemes of interaction of stress waves:
a — in series; b — in parallel.

Let us study the spectral energy for a better and complete assessment of the interaction of blast pulses. Using formula (2.5) for the above schemes, we get the following corresponding expressions of energy distribution as per frequency.

$$\frac{dE_{S_1}}{d\omega} = N_1(\omega) = \frac{4\sigma_{max}^2}{\pi} \frac{\sin^2 \omega\tau}{\omega^2};$$

$$\frac{dE_{S_2}}{d\omega} = N_2(\omega) = \frac{16\sigma_{max}^2}{\pi} \frac{\sin^2 \omega\tau/2}{\omega^2}.$$

The energy distribution curves versus frequency are shown in Fig. 5. A comparison of the above relationships shows that for the first scheme (curve 1) the main frequency of blast pulse energy is distributed at lower frequencies (1–20 Hz), while for the second scheme (curve 2), it is shifted to the region of higher frequencies (1–50 Hz).

From the frequency diagram it is evident that as regards the compaction of subsided loess massifs one can achieve the best effect by making the stress waves interact in series, as the wave disturbances at lower frequencies attenuate less.

Thus, a study of the amplitude-frequency spectrum of blast effects permits us to discern the patterns of wave propagation as well as to develop more effective technological schemes for blasting operations using dynamic loads of various spectral compositions. In particular, based on the sequence of interaction of blast waves of various intensities on the ground, soil compaction can be controlled at a distance from the charge within the near-field zone or in the zone of microdeformations. These conclusions can further be used in engineering problems for enhancing the validity and accuracy of calculations in designs.

Dynamics of formation of amplitude-frequency spectrum: The amplitude-frequency spectrum of a wavefield is a very important characteristic of any blasting process. Using this characteristic, we can study in depth the process of compaction with the help of blast energy, the seismic effects on buildings and

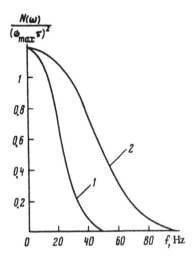

Fig. 5: Curves showing energy distribution $N(\omega)$ versus f.

other phenomena. To develop effective methods of controlling the blast pulse, one should consider the amplitude-frequency spectrum over time.

Let the diagram of stress waves along the radial component be described by [11]

$$\sigma_r(r, t) = \sigma_r^{max}(r) \exp[-\theta(t - t_{max})] \frac{\sin\beta(t - t_0)}{\sin\beta(t_{max} - t_0)}. \qquad \ldots (2.11)$$

The stress wave represented by formula (2.11) corresponds to the actual diagram (Fig. 6). In formula (2.11) the following notations are used:

σ_r^{max} is the coefficient of maximum amplitude of stress wave, varying with time (A.A. Vovk, G.I. Chernyi and A.G. Smirnov. *Deformation of Compressible Media under Dynamic Loads*, Naukova Dumka, Kiev, 1971);

$$\sigma_r^{max} = k_\sigma r_0^{-\mu_\sigma}, \qquad \ldots (2.12)$$

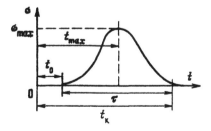

Fig. 6: Diagram of stress wave.

where k_σ, μ_σ are experimental coefficients.

θ is the coefficient characterising the curvature of rise and fall of amplitude of stresses over time:

$$\theta r_c = a_1 + a_2 r_0. \qquad \ldots (2.13)$$

Time of arrival of the first disturbance at a given point is given by

$$t_0/r_c = b_1 + b_2 r_0. \qquad \ldots (2.14)$$

Time of rise of amplitude of stress to its maximum value

$$t_{max}/r_c = d_1 r_0^{d_2}. \qquad \ldots (2.15)$$

Coefficient $\beta = \pi/\tau$ characterises the duration of positive phase of stresses (τ is the duration of positive phase of the blast pulse and varies as per $\tau/r_c = c_1 + c_2 r_0$).

$$\beta r_c = f_1 + f_2 r_0. \qquad \ldots (2.16)$$

Let us define the complex amplitude-frequency spectrum of the stress wave given in the form of (2.11),

$$S(j\omega) = \int_{t_0}^{t_k} \sigma_r^{max} e^{-\theta(t-t_{max})} \frac{\sin\beta(t - t_0)}{\sin\beta(t_{max} - t_0)} e^{-j\omega t} dt, \qquad \ldots (2.17)$$

where t_0, t_k are initial and final time of action of blast disturbance respectively.

Let us rewrite (2.17) in the form

$$S(j\omega) = \frac{\sigma_r^{max} e^{\theta t_{max} t}}{\sin\beta(t_{max} - t_0)} \int_{t_0}^{t_k} e^{-\kappa t} \sin\beta(t - t_0) dt,$$

where $\kappa = \theta + j\omega$.

Denoting $I = \int_{t_0}^{t_k} e^{-\kappa t} \sin\beta(t - t_0) dt$

then in an expanded form

$$I = \frac{\kappa \sin\beta(t - t_0) e^{-\kappa t} - \beta e^{-\kappa t} \cos\beta(t - t_0)}{\kappa^2 + \beta^2}$$

$$S(j\omega) = \frac{\sigma_r^{max} e^{\theta t_{max}}}{\sin\beta(t_{max} - t_0)} \left[\frac{-\kappa e^{-\kappa t}}{\kappa^2 + \beta^2} \sin\beta(t - t_0) - \frac{\beta e^{-\kappa t}}{\kappa^2 + \beta^2} \cos\beta(t - t_0) \right]_{t_0}^{t_k} =$$

$$= \frac{\sigma_r^{\max} e^{\theta t_{\max}}}{\sin \beta (t_{\max} - t_0)} \left(\frac{\beta e^{-\kappa t_0} + \beta e^{-\kappa t_k}}{\kappa^2 + \beta^2} \right). \qquad \ldots (2.18)$$

Thus the required complex spectrum, by considering the introduced notation for κ, is obtained as

$$S(j\omega) = \frac{\beta \sigma_r^{\max} e^{\theta t_{\max}}}{[(\theta + j\omega)^2 + \beta^2] \sin \beta (t_{\max} - t_0)} \times$$

$$\times \left[e^{-(\theta + j\omega)t_0} + e^{-(\theta + j\omega)t_k} \right]. \qquad \ldots (2.19)$$

As the amplitude-frequency spectrum does not depend on the selected initial reference time (A.A. Kharkevich. *Spectra and Analysis*. Fizmathgiz, Moscow, 1962), to simplify formula (2.19) we may substitute $t_0 = 0$ for each distance. In other words, let us bring all stress pulses at some distance from the origin of time frame $t_0 = 0$. We shall find in this case the amplitude-frequency spectrum as the modulus of complex spectrum $S(j\omega)$:

$$F(\omega) = |S(j\omega)| = \frac{\beta \sigma_r^{\max} e^{-\theta(\tau - t_{\max})}}{\sin \beta t_{\max}} \times$$

$$\times \sqrt{\frac{1 + e^{2\theta\tau} + 2e^{\theta\tau} \cos \omega\tau}{(\theta^2 + \beta^2 - \omega^2) + 4\theta^2 + \omega^2}}. \qquad \ldots (2.20)$$

This expression is obtained assuming that θ is independent of time, i.e., is a constant quantity for the stress wave. Actually θ varies with time and is determined by the formula

$$\theta(t) = Y / \log e^{(t_{\max} - t)}, \qquad \ldots (2.21)$$

where $Y = \log \sigma_r - \log \sigma_r^{\max} - \log \dfrac{\sin \beta t}{\sin \beta t_{\max}}$.

The amplitude-frequency spectrum (2.20) is obtained under the condition that the upper limit of the integral in (2.17) is equal to $t = t_K$. If this limit is variable, then the expression for the amplitude-frequency spectrum is obtained in the form

$$F(\omega, t) = \frac{\sigma_r^{\max} e^{-\theta(t - t_{\max})}}{\sin \beta t_{\max}} \times$$

$$\times \sqrt{\frac{(\beta e^{-\theta t} \cos \omega t - \beta \cos \beta t - \theta \sin \beta t)^2 + (\beta e^{\theta t} \sin \omega t - \omega \sin \beta t)^2}{(\theta^2 + \beta^2 - \omega^2)^2 + 4\theta^2 \omega^2}}$$

$$\ldots (2.22)$$

Expression (2.22) enables us to study the dynamics and nature of formation of amplitude-frequency spectrum related to time. From this expression at $t = t_K$, we get relationship (2.20).

Thus the amplitude-frequency spectrum depends on the stress wave parameters which are functionally determined by the relationships (2.12) to (2.16).

To find the relationships (2.12) to (2.16) for loess-like loamy sands, experiments were conducted to study the effect of blasting a single cylindrical charge DSh-A weighing $C_1 = 0.05$ kg per metre ($r_c = 0.003$ m). As a result of experimental investigations, for distances (60–140) r_c the following values of coefficients contained in formulae (2.12) to (2.16) were obtained:

$$
\begin{aligned}
a_1 &= 5.75; & a_2 &= -0.036 \text{ m/s}; \\
b_1 &= -0.3; & b_2 &= -8 \cdot 10^{-3} \text{ m/s}; \\
d_1 &= 2.2 \cdot 10^{-4} \text{ s/m}; & d_2 &= 1.81; \\
f_1 &= 1.53; & f_2 &= -6.4 \cdot 10^{-3} \text{ m/s}; \\
c_1 &= 1; & c_2 &= 0.028; \\
k_\sigma &= 10 \text{ GPa}; & \mu_\sigma &= 2.17.
\end{aligned}
$$

At $r_c = 0.003$ m, the initial data for calculating the amplitude-frequency spectrum of stress waves of the type (2.11) are given in Table 4. The curves of the amplitude-frequency spectrum at various time periods are shown in Fig. 7.

The curves of the amplitude-frequency spectrum versus time are shown in Fig. 8. It is evident that at the initial moment of time ($t = 0.3$ ms), components of the spectrum in the range 0–400 Hz practically do not vary in amplitude. With time, the amplitude of higher frequency components 100–400 Hz decreases sharply. In the 0–50 Hz range, the amplitude-frequency spectrum varies insignificantly.

Variation in the amplitude-frequency spectrum at various distances is shown in Fig. 9. The nature of these curves confirms the conclusions about the attenuation of spectrum with increasing distance.

Figure 10 illustrates the relationship between the amplitude-frequency spectrum and time t_{max} at which the stress wave attains its maximum value. The amplitude of frequency spectrum reduces as t_{max} increases. In the 0–50 Hz range, a weak attenuation of amplitude of spectrum is observed, which confirms

Table 4

r_0	$\sigma_r^{max} \times 10^{-4}$, Pa	β, s^{-1}	θ, s^{-1}	t_0, s	t_{max}, s	τ, s
60	138	383	1196.6	0.00054	0.00108	0.0081
80	74	339.3	956.6	0.0010	0.0018	0.0098
100	46	296.6	716.6	0.0015	0.0027	0.0105
120	31	254	476.6	0.0020	0.0038	0.0132
140	22	211.3	236.6	0.0025	0.0051	0.0149

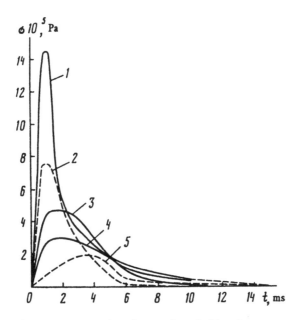

Fig. 7: Diagrams of stress waves at various distances from the blast site:
$1-60\ r_c$; $2-80\ r_c$; $3-100\ r_c$; $4-120\ r_c$; $5-140\ r_c$.

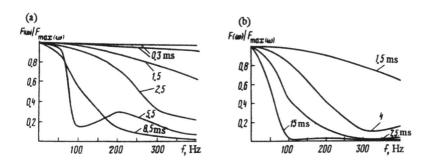

Fig. 8: Curves of amplitude-frequency spectrum of different time periods at relative distances
(a) $r = 60\ r_c$ and (b) $r = 40\ r_c$.

insignificant losses in the energy of blast pulses at these frequencies. At a frequency exceeding 50 Hz, the amplitude of the spectrum decreases more sharply compared to the amplitude in the 0–50 Hz interval.

Variation in certain components of the spectrum over time is shown in Fig. 11. These curves illustrate the dynamics of formation of the amplitude spectrum.

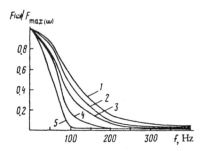

Fig. 9: Curves of amplitude-frequency spectrum at different distances:
$1-60\ r_c$; $2-80\ r_c$; $3-100\ r_c$; $4-120\ r_c$; $5-140\ r_c$.

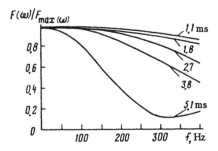

Fig. 10: Curves of amplitude-frequency spectrum depending on time t_{max}.

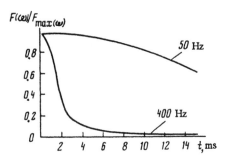

Fig. 11: Variation in components of amplitude-frequency spectrum over time.

The dependence of amplitudes of components in the frequency spectrum on the duration of action of the positive phase of blast pulse β is shown in Fig. 12,a. As β increases, an increase in the amplitude of all components in the frequency spectrum is observed. For components of 50 and 100 Hz the occasional fall of amplitudes is more than that for the components of 300 and 400 Hz. For 50 Hz components in the spectrum, at $\log \theta$ values equal to 2.15–2.9, a decrease

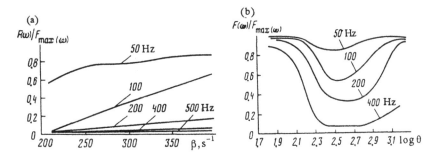

Fig. 12: Relationship between components of amplitude-frequency spectrum and the parameters β and θ.

in amplitude is noticed (Fig. 12,b). For 100 Hz component, the range of $\log \theta$ values at which fall in reduction of amplitude is noticeable broadens from 2.15 to 3.1. The maximum fall in amplitude of the spectrum is attained for the 400 Hz component.

Thus the amplitude-frequency spectrum depends essentially on the values of coefficient θ. There is a specific interval of variations of this coefficient, at which amplitude of the spectrum decreases markedly and particularly for components of the spectrum up to 50 Hz this interval is smaller than for components exceeding 50 Hz.

The data obtained could be used for creating a blast pulse with a given set of parameters that would allow specific technological problems to be solved.

2.3 Frequency Spectrum of Seismic Vibrations due to Blasting of a Spherical Charge

Several experimental investigations were carried out to study the conditions for initiating seismic waves. Various theoretical studies were done to explain the observed effects. The theory treating the source as a spherical cavity is widely accepted. A model of blast-induced elastic wave propagation based on this theory has been suggested [30] and the concept of equivalent radius of the source is introduced. However, the analysis is limited to some types of function $\bar{p}(t)$, describing the external effect, assuming that the constants of medium $\lambda = \mu$ (λ and μ, Lame's constants). In later developments of the theory of a spherical source, attention was mainly paid to the study of the processes taking place inside the spherical cavity and to the clarification of the type of function $p(t)$. The possibility of explaining the observed phenomena through the theory of spherical source remained barely explored.

Let us consider the problem of the velocity spectrum of a seismic wave in the following formulation (problem solved in collaboration with V.G. Kravtsev). Let a source of seismic vibrations with radius r_e, be working in an absolute elastic isotropic medium, characterised by the elastic constants of Lame λ and μ and density ρ. At the boundary of the source, the applied action is

$$\bar{p}(t) = \sigma(t). \qquad \ldots (2.23)$$

The field of disturbances arising from such action has a spherical symmetry and satisfies the wave equation

$$\frac{\partial^2 a}{\partial r^2} + \frac{2}{r}\frac{\partial a}{\partial r} - \frac{2a}{r^2} = \frac{1}{v_P^2}\frac{\partial^2 a}{\partial t^2}, \qquad \ldots (2.24)$$

where $v_P^2 = \dfrac{\lambda + 2\mu}{\rho}$, square of velocity of longitudinal waves.

The solution to (2.24) can be found in:

$$a = \frac{\partial f(\tau/r)}{\partial r} = -\frac{f(\tau/r)}{r^2} - \frac{1}{rv_P}\frac{\partial f(\tau/r)}{\partial \tau}, \qquad \ldots (2.25)$$

where $\tau = t - \dfrac{r - r_e}{v_P}$;

$f(\tau/r)$ is an unknown function which can be established from boundary conditions:

$$\left[(\lambda + 2\mu)\frac{\partial a}{\partial r} + 2\lambda\frac{a}{r}\right]_{r=r_e} = -\sigma(t). \qquad \ldots (2.26)$$

From (2.25), we have

$$\frac{\partial a}{\partial r} = \frac{1}{rv_P^2}\frac{\partial^2 f(\tau/r)}{\partial \tau^2} + \frac{2}{r^2 v_P}\frac{\partial f(\tau/r)}{\partial \tau} + \frac{2f(\tau/r)}{r^3}. \qquad \ldots (2.27)$$

Substituting in the boundary conditions (2.26) the expression for a, we get as per (2.25) and (2.27)

$$\sigma(t) = -\frac{\lambda + 2\mu}{r_e v_P^2} + f''(\tau/r) - \frac{4\mu}{r_e^2 v_P}f'(\tau/r) - \frac{4\mu}{r_e^3}f(\tau/r). \qquad \ldots (2.28)$$

Multiplying the RHS and LHS of (2.28) by $e^{-j\omega t}$ and integrating it over ω from $-\infty$ to $+\infty$, we get

$$S_\sigma(j\omega) = -\frac{\lambda + 2\mu}{r_e v_P^2}(j\omega)^2 S_f(j\omega) - \frac{4\mu}{r_e^2 v_P}j\omega S_f(j\omega) - \frac{4\mu}{r_e^3}S_f(j\omega) =$$

$$= S_f(j\omega)\left(\frac{4\mu}{r_e^3} + \frac{4\mu j\omega}{r_e^2 v_P} - \frac{\lambda + 2\mu}{r_e v_P^2}\omega^2\right), \qquad \ldots (2.29)$$

where $S_\sigma(j\omega)$ is the complex spectrum of function $\sigma(t)$;

$S_f(j\omega)$ is the complex spectrum of function $f(\tau/r)$ to be found.

Let us find the complex spectrum of displacement $S_a(j\omega)$. For this, let us take recourse to the Fourier transformation of (2.25).

$$S_a(j\omega) = -S_f(j\omega)/r^2 - j\omega S_f(j\omega)/(rv_P) =$$
$$= -S_f(j\omega)\left[1/r^2 + j\omega/(rv_P)\right]. \qquad \ldots (2.30)$$

From (2.30), the spectrum of unknown function $f(\tau/r)$ can be found out

$$S_f(j\omega) = -S_a(j\omega)/[1/r^2 + j\omega/(rv_P)]. \qquad \ldots (2.31)$$

Substituting (2.31) in (2.29), we get

$$S_\sigma(j\omega) = \frac{S_a(j\omega)}{1/r^2 + j\omega/(rv_P)}\left[\frac{4\mu}{r_e^3} + \frac{4\mu j\omega}{r_e^2 v_P} - \frac{(\lambda + 2\mu)\omega^2}{r_e v_P^2}\right]. \qquad \ldots (2.32)$$

From (2.32) the complex spectrum of displacements generated by the action of $\sigma(t)$ at the seismic source boundary,

$$S_a(j\omega) = \frac{r_e[1/r^2 + j\omega/(rv_P)]}{4\mu/r_e^2 - (\lambda + 2\mu)\omega^2/v_P^2 + 4\mu j\omega/(r_e v_P)}S_\sigma(j\omega). \qquad \ldots (2.33)$$

Denoting

$$M(j\omega) = \frac{r_e[1/r^2 + j\omega/(rv_P)]}{4\mu/r_e^2 - (\lambda + 2\mu)\omega^2/v_P^2 + 4\mu j\omega/(r_e v_P)}. \qquad \ldots (2.34)$$

Let us rewrite (2.33) in the following form,

$$S_a(j\omega) = S_\sigma(j\omega)M(j\omega). \qquad \ldots (2.35)$$

According to [13], $M(j\omega)$ can be considered a complex frequency spectrum of a spherical emitter of longitudinal waves at distances $r > r_e$. The relationship (2.35) can be physically traced as follows. The action of a spherical emitter relative to the action of any other equivalent filter can be analysed by considering pressure $p(t)$ as an input function of the emitter's oscillation system, while displacement $a(t)$ is considered as an output function. I.I. Gurvich determined the frequency characteristics and transition functions of the emitter and their dependence on charge mass and properties of the medium, by establishing an analogy between an emitter and a resonance filter.

As the recording instruments used for registering vibrations in the ground usually record only displacement velocities, it is advisable to determine the spectrum of displacement velocities $S_e(j\omega)$.

To find the spectrum $S_e(j\omega)$, let us multiply LHS and RHS of the equality (2.33) by $j\omega$. Then

$$S_e(j\omega) = j\omega S_a(j\omega) = \frac{j\omega r_e[1/r^2 + j\omega/(rv_P)]S_\sigma(j\omega)}{4\mu/r_e^2 - (\lambda + 2\mu)\omega^2/v_P^2 + 4\mu j\omega/(r_e v_P)}$$

or, after simple transformations,

$$S_e(j\omega) = \frac{S_\sigma(j\omega)(j\omega v_P^2 - \omega^2 r v_P)r_e}{4\mu r^2[(\lambda + 2\mu)\omega^2/(4\mu) - (v_P/r_e)^2 - j\omega(v_P/r_e)]}. \qquad \ldots (2.36)$$

From (2.36) we shall find the amplitude frequency spectrum of displacement velocities:

$$F(\omega) = |S_e(j\omega)| =$$

$$= \frac{|S_\sigma(j\omega)|r_e v_P\omega\sqrt{v_P + r^2\omega^2}}{4\mu r^2\sqrt{(v_P/r_e)^4 + [1 - (\lambda + 2\mu)/(2\mu)](v_P/r_e)^2\omega^2 + [(\lambda + 2\mu)/(4\mu)]^2\omega^4}}.$$

$$\ldots (2.37)$$

In order to find the spectrum showing the effect of $\sigma(t)$ in formula (2.37), let us consider the case when at the boundary of seismic wave source a blast pulse of type (2.8) acts. The spectral characteristic of this pulse is determined by formula (2.10). Thus, taking into account (2.10), the expression (2.37) takes the following form:

$$F(\omega) = \sigma_{max} v_P r_e/(4\mu r^2 a_e b_e \omega\tau)[1 + a_e^2 b_e^2 + 2a_e b_e \cos\omega\tau -$$

$$- 2(a_e \cos b_e\omega\tau + b_e \cos a_e\omega\tau)]^{1/2} \times$$

$$\times (v_P^2 + r^2\omega^2)^{\frac{1}{2}} [(v_P/r_e)^4 + [1 - (\lambda + 2\mu)/(2\mu)](v_P/r_e)^2\omega^2 +$$

$$+ [(\lambda + 2\mu)/(4\mu)]^2\omega^4]^{-\frac{1}{2}}. \qquad \ldots (2.38)$$

From an analysis of (2.38) it follows that the spectrum of displacement velocities $F(\omega)$ depends not only on the parameters of blast effect a_e, b_e, τ but also on the rock massif parameters λ, μ, v_P, r_e.

Effect of Physicomechanical Properties of Medium on Variation in Frequency Spectrum of Seismic Waves: Before switching over to the study of frequency spectrum determined by the general relationship (2.38), let us consider the method of establishing the radius of emitter r_e. According to F.F. Aptikaev, by the radius of emitter is meant that area whose boundary can be determined by characteristic indicators of waves (change in the pattern of seismic wave attenuation), as can be seen in seismograms. The boundary of this area corresponds to the boundary between inelastic and elastic zones.

There are several contradictions in the literature regarding the zonal dimensions of the source (area of elastic and inelastic zones). As per Yu.V. Bondarenko, the dimensions of this zone on average are $r_e = 9\sqrt[3]{C} \approx 166 \, r_c$. According to B.G. Rulev, the boundary of the source zone $r_e = 2.5\sqrt[3]{C} \approx 46 \, r_c$.

While determining the source radius, we shall rely upon the relationships given in [3],

$$\log r_e = -1.2 + 0.4 \log E_e,$$

where E_e is the quantum of seismic energy at the boundary of emitter, J. Then,

$$r_e = 0.063 \, E_e^{0.4}. \qquad \ldots (2.39)$$

Based on a large volume of experimental data, F.F. Aptikaev assessed the influence of the type of explosive, depth of charge placement, certain physicomechanical properties of the medium and its moisture content. The seismic energy of a blast is expressed as:

$$E_e = k k_{\text{expl}} k_m k_w k_z C, \qquad \ldots (2.40)$$

where C is the charge mass;

k is the proportionality coefficient;

k_{expl} is the coefficient that takes into account the type of explosive;

k_m is the coefficient that takes into account the physicomechanical properties of the medium;

k_w is the coefficient that considers the moisture content of the test sample;

k_z is the coefficient considering the depth of placement of the explosive charge.

If the energy of a charge is evaluated by its mass, then it is also essential to take into account the quality of the explosive; k_{expl} is considered to be proportional to the specific energy of the explosive.

The coefficients k_m for all media except for underwater blasting are approximately equal and their values can be included in the proportionality coefficient k. The coefficient k_w takes into account the physicomechanical properties of a medium to some extent as the moisture in a medium is correlated to its other parameters, such as porosity, density etc. Underwater explosives can conveniently be considered an extreme case while studying the effect of moisture in a medium on seismic effectiveness.

It has been observed that the dimensions of blasting cones and amplitudes of seismic signals correlate well with the moisture content in a medium. To establish the relationship between k_w and the dimensions of blasting cones, the following approach was used. According to published data [3], a relationship between the radii of blasting cones and the charge mass, taking into account the coefficient k_{expl}, was derived. Coefficient k_z was not considered as the dimensions of cones were related to the case of optimal depth of charge placement. Based on these data, a curve showing the relationship between the coefficient k_w and moisture content in the medium was drawn.

F.F. Aptikaev obtained a graphic relationship between the coefficient k_z and scaled distance of charge placement.

Substituting in (2.40) the values of seismic energy of blasts calculated as per seismograms and using the obtained values of k_{expl}, k_w, k_z, the coefficient of proportionality k was found to be $5 \cdot 10^4$. Formula (2.40) can now be transformed:

$$E_e = 5 \cdot 10^4 k_{expl} k_z k_w C. \qquad \ldots (2.41)$$

For a charge of trotyl causing the camouflet effect in a medium with moisture $w = 10-20\%$, F.F. Aptikaev obtained the following values for the coefficients: $k_{expl} = 10, k_z = 12, k_w = 0.12$. Then as per formula (2.39) using relationship (2.41) we obtain

$$r_e = 5.52\, C^{0.4}. \qquad \ldots (2.42)$$

Let us consider the case when at the boundary of seismic source an action of type (2.8) is applied having parameters $\tau_1 = 0.01$ s, $\tau_2 = 0.06$ s. Let the rock massif contain sand of $\rho = 1.75$ T/m^3, the velocity of longitudinal waves $v_P = 700$ m/s and the velocity of transverse waves $v_S = 380$ m/s. Using the known relationships between elastic characteristics, we get $\lambda = 3521 \cdot 10^5$ Pa, $\mu = 2527 \cdot 10^5$ Pa, $\nu = 0.29$.

While studying the spectrum of seismic vibrations induced by blasting a spherical charge, three cases were considered [13]: $r_e = 5.83$ m, $r_e = 10$ m, $r_e = 35$ m. After substituting these constants in (2.38), the following formulae were obtained for each case respectively:

$$F_1(\omega) = S_\sigma(\omega) \frac{0.0018\omega(225\omega^2 + 490000)^{\frac{1}{2}}}{(20736 \cdot 10^4 - 10368\omega^2 + 0.72\omega^4)^{\frac{1}{2}}};$$

$$F_2(\omega) = S_\sigma(\omega) \frac{0.0031\omega(225\omega^2 + 490000)^{\frac{1}{2}}}{(2401 \cdot 10^4 - 3430\omega^2 + 0.72\omega^4)^{\frac{1}{2}}}; \qquad \ldots (2.43)$$

$$F_3(\omega) = S_\sigma(\omega) \frac{0.0107\omega(225\omega^2 + 490000)^{\frac{1}{2}}}{(16 \cdot 10^4 - 280\omega^2 + 0.72\omega^4)^{\frac{1}{2}}};$$

where $S_\sigma(\omega) = \dfrac{\sigma_{max} 10^{-5}}{0.00846\omega}[1.72 + 0.28\cos 0.06\omega - 2(0.17\cos 0.05\omega + 0.83\cos 0.01\omega)]^{\frac{1}{2}}.$

The results of these calculations are shown in Fig. 13. The ratio between current value of the amplitude of spectral density to its maximal value is plotted on axis y. From an analysis of these curves it follows that the dimensions of emitter r_e significantly affect the frequency composition of the spectrum as well

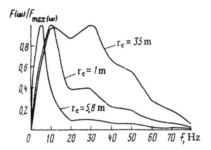

Fig. 13: Dependence of the spectrum of seismic vibrations on r_e and frequency f.

as the values of frequencies at which the spectrum of seismic vibrations attains the maximum value.

The relationship $f_{max} = 44.66 r_e^{-0.625}$ follows from the curves shown in Fig. 13. The curve showing variation in f_{max} due to r_e is shown in Fig. 14a. Its nature does not contradict the experimental data obtained by A.A. Kharkevich.

Another relationship can be obtained from Fig. 13: $f_{max} = 3 + 0.1 v_P / r_e$. This is shown in Fig. 14,b.

Thus, as the ratio v_P / r_e is increased, the maximum of the seismic vibrations spectrum shifts to the region of higher frequencies. The velocity of longitudinal waves depends largely on the physicomechanical properties of rocks and particularly their moisture. The radius of emitter depends both on the properties of the medium and the type of explosive used.

Effect of Short Delay Blasting (SDB) on Variation in Frequency Spectrum of Seismic Vibrations: Use of SDB enables one to obtain desired technological properties of a rock massif: uniform fragmentation during breakage and optimal density during compaction of massifs containing considerable soil. In this context, studying the effect of SDB on the spectrum of seismic vibrations is of considerable importance.

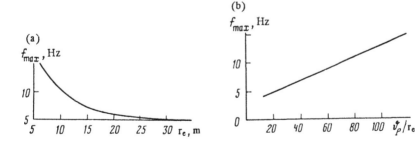

Fig. 14: Dependence of f_{max} on (a) radius of emitter, r_e and (b) the ratio v_P / r_e.

Let us select the law of change in displacement velocity of seismic disturbance in the following manner:

$$u(t) = u_{max}e^{-\beta_u t}\sin\omega_0 t. \qquad \dots (2.44)$$

We shall use the known expression of spectral analysis for summing two pulses shifted in time by $\Delta\tau$ from each other, as obtained by F.F. Aptikaev, for calculating the velocity of blast-induced seismic vibrations,

$$S_{\Delta\tau}(j\omega) = S_0(j\omega) + S_0(j\omega)e^{j\omega}\Delta\tau,$$

where $S_0(j\omega)$ is the complex spectrum of unit pulse (2.44);

$\Delta\tau$ is the delay interval.

The complex spectrum of the resultant seismic vibrations, when several unit pulses with delay interval $\Delta\tau$ are summed up, is given by the sum

$$S_{\Sigma}(j\omega) = S_0(j\omega) + S_0(j\omega)e^{j\omega\tau} + \cdots + S_0(j\omega)e^{mj\omega\Delta\tau}, \qquad \dots (2.45)$$

where m is the number of groups being blasted.

The complex spectrum $S_{\Sigma}(j\omega)$ can be defined as the sum of the geometrical progression series with numerator $e^{j\omega\Delta\tau}$.

$$S_{\Sigma}(j\omega) = S_0(j\omega)\frac{1 - e^{(m+1)j\omega\Delta\tau}}{1 - e^{j\omega\Delta\tau}}.$$

Then the amplitude spectrum

$$F_{\Sigma}(\omega) = |S_0(j\omega)|\frac{|1 - e^{(m+1)j\omega\Delta\tau}|}{|1 - e^{j\omega\Delta\tau}|};$$

$$F_{\Sigma}(\omega) = |S_0(j\omega)|\left[\frac{\left(1 - e^{(m+1)j\omega\Delta\tau}\right)\left(1 - e^{-(m+1)j\omega\Delta\tau}\right)}{\left(1 - e^{-j\omega\Delta\tau}\right)\left(1 - e^{-j\omega\Delta\tau}\right)}\right]^{\frac{1}{2}} =$$

$$= |S_0(j\omega)|[(1 - \cos(m + 1)\omega\Delta\tau) / (1 - \cos\omega\Delta\tau)]^{\frac{1}{2}}.$$

After simple trigonometric transformations, we get

$$F_{\Sigma}(\omega) = |S_0(j\omega)|\frac{\sin(m + 1)\omega\Delta\tau/2}{\sin\omega\Delta\tau/2}. \qquad \dots (2.46)$$

The amplitude-frequency spectrum of unit pulse (2.44) is given by

$$|S_0(j\omega)| = \frac{u_{max}}{\omega_0\left\{[1 + (\beta_u/\omega_0)^2 - (\omega/\omega_0)^2]^2 + 4(\beta_u/\omega_0)^2(\omega/\omega^2)\right\}^{\frac{1}{2}}}.$$

53

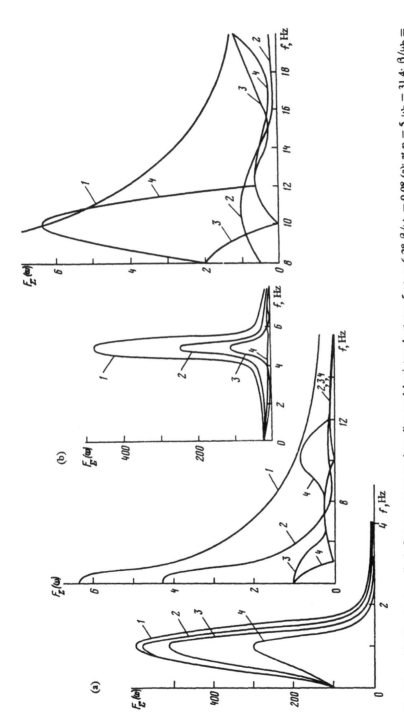

Fig. 15,a and b: Resultant amplitude-frequency spectrum depending on delay intervals at $n = 5$, $\omega_0 = 6.28$, $\beta/\omega_0 = 0.08$ (a); at $n = 5$, $\omega_0 = 31.4$; $\beta/\omega_0 = 0.02$ (b).

1 — 0 ms; 2 — 20 ms; 3 — 50 ms; 4 — 100 ms.

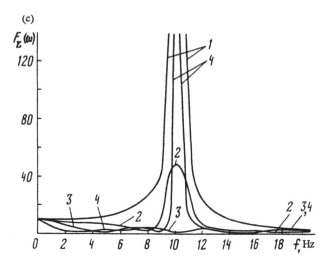

(c)

Fig. 15c: where $n = 5, \omega_0 = 62.8, \beta/\omega_0 = 0.016$.

Thus the resultant spectrum of seismic vibrations is determined by the expression

$$F_\Sigma(\omega) = \frac{u_{max} \sin(m+1)\omega\Delta\tau/2}{\omega_0 \sin \dfrac{\omega\Delta\tau}{2} \left\{[1 + (\beta_u/\omega_0)^2 - (\omega/\omega_0)^2]^2 + 4(\beta_u/\omega_0)^2(\omega/\omega_0)^2\right\}^{\frac{1}{2}}}. \quad \ldots (2.47)$$

When $m + 1$ charges are blasted simultaneously ($\Delta\tau = 0$), from (2.47) we obtain the resultant spectrum,

$$F_{\Sigma simul}(\omega) = \frac{(m+1)u_{max}}{\omega_0 \left\{[1 + (\beta_u/\omega_0)^2 - (\omega/\omega_0)^2]^2 + 4(\beta_u/\omega_0)^2(\omega/\omega_0)^2\right\}^{\frac{1}{2}}}. \quad \ldots (2.48)$$

From (2.47) and (2.48) it is evident that at $\omega = \omega_0$ maximum values of $F_\Sigma(\omega)$ and $F_{\Sigma simul}(\omega)$ are obtained.

While studying the spectrum of seismic vibrations, the following parameters were selected: β_u, coefficient characterising the attenuation of displacement velocity of seismic disturbance in 1–0.5 interval; natural frequency of seismic disturbance in the interval $f_0 = 0.8$–16 Hz ($\omega_0 = 5$–100). The parameter β_u/ω_0 varied in the interval (0.2–0.01).

An analysis of these investigations of the seismic spectrum related to the SDB of five charges with different delay intervals revealed that the amplitude-frequency spectra depended largely on the natural frequency of seismic disturbance ω_0. Curves of spectral density at $\omega_0 = 6.28, f_0 = 1$ Hz (Fig. 15,a), $\omega_0 = 31.4, f_0 = 5$ Hz (Fig. 15,b), $\omega_0 = 62.8, f_0 = 10$ Hz (Fig. 15,c) are shown in Fig. 15. In the region of frequencies nearer to ω_0, variation in the

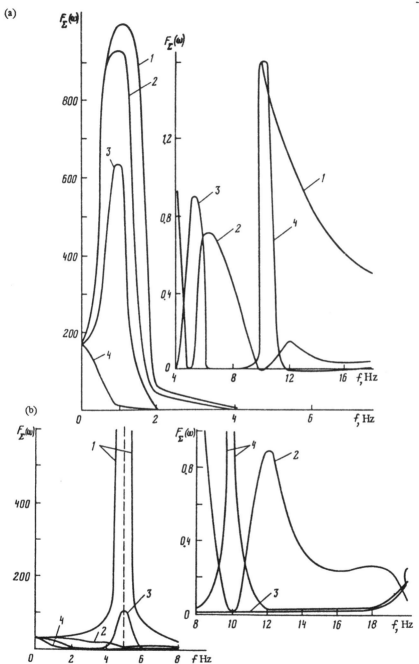

Fig. 16a and b: Amplitude-frequency spectrum depending on the delay interval, where $n = 9, \omega_0 = 6.28; \beta/\omega_0 = 0.08$ (a); $n = 9, \omega_0 = 31.4; \beta/\omega_0 = 0.02$ (b).

1—0 ms; 2—20 ms; 3—50 ms; 4—100 ms.

56

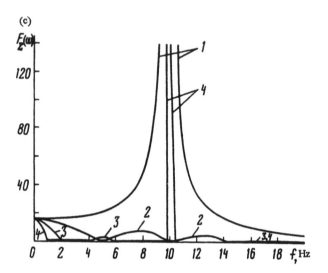

Fig. 16c: where $n = 9, \omega_0 = 62.8, \beta/\omega_0 = 0.016$.

amplitude of frequency spectrum exhibited a resonance character and the delay interval in particular affected the amplitude of the spectrum. The influence of delay interval was not identical in all five cases. For example, the delay interval $\Delta\tau = 100$ ms for the natural frequency $\omega_0 = 62.8$ (Fig. 15) produced a resonance effect on the nature of variation in the amplitude of spectrum within the frequency region nearer to 10 Hz. On the right-hand side of the figures, curves of amplitude-frequency spectra are shown in a stepped-up time scale.

The amplitude frequency spectrum of a blast of 10 charges with different delay intervals is shown in Fig. 16. The nature of variation in these curves corresponds to the results obtained above for $m = 5$. It can be seen that the number of groups blasted in the cases considered increased the amplitude of spectrum.

The influence of delay interval $\Delta\tau$ on the amplitude of certain components in the frequency spectrum is depicted in Fig. 17. The nature of variation of three spectral components depending on $\Delta\tau$ at $m = 9$, is drawn in the diagram. It is evident from the figure that the delay interval exerted considerable effect on amplitude only in a particular range. There are ranges in which the delay interval did not contribute to the increase in amplitude of the spectrum. Depending on the class of problems being solved and on the basis of the data obtained, one can select a delay interval rationally.

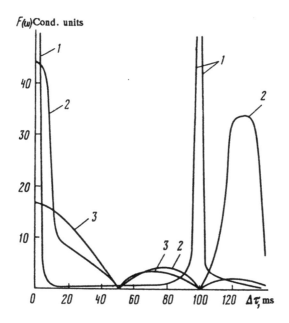

Fig. 17: Relationship between amplitude of frequency spectrum F and delay interval $\Delta\tau$ at $n = 9, \omega_0 = 62.8, \beta/\omega_0 = 0.016$: 1—at 10 Hz; 2—8 Hz; 3—2 Hz.

3

Effect of Blasting a Cylindrical Charge in Soils with a Free Face

3.1 Characteristics of Near- and Far-field Blast Effects

It is important to know how and where the waves generated by a blast form in order to optimally utilise their energy for creative purposes in certain cases to vary the properties of soils by expanding the parameters of microdeformation zones, and in other cases to effectively counter the blast effect. This can be achieved by an in-depth study of all the aspects and dimensions of the area subjected to blast. Studies on the effects induced by blasting concentrated charges on a rock massif with a free face have shown that such a blast can, and ought to be considered as having occurred in two phases, i.e., as a two-phase source of vibrations. The first source, the initial phase in development of a blast, when its products expand symmetrically in all directions, results in compressive deformations in the surrounding soil which generate body wave P. The second source, a dome-shaped heaving of soil in the epicentral zone, is induced by residual pressure of gases in the cavity and the associated low frequency inelastic vibrations in the near-field zone.

Research into the aspects of blasting with spherical and cylindrical charges has shown that there are no principal differences in the wave spectra within the trajectories of particle movement. Therefore, it can be assumed that blasting of a cylindrical charge can also be explained as having occurred in two phases. This assumption is further confirmed by a series of factors obtained from the study of vibrations in soil induced by blasting cylindrical charges.

In the study of the trajectories of particle movement, it was observed that the first blast waves to arrive were oriented towards the blasting source and that low-frequency vibrations were superposed over high-frequency ones among the initial P-wave arrivals. Furthermore, these low-frequency vibrations appeared on the oscilloscope at that point of time when the wave compression had already reflected back from the free surface. Thus at the free surface a characteristic picture of ground movement at the blast's epicentre, having two maxima of movement, P and R_1, among the initial waves to arrive (Fig. 18) emerged.

The surfacing of components of these low-frequency ground movements (called R) took place at the same time interval within the range of relative dis-

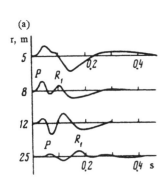

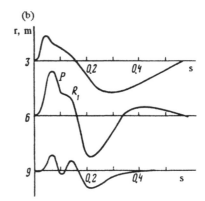

Fig. 18: Characteristic oscillograms of surface ground movements in the near-field zone of a blast: a — vertical cylindrical charge ($C = 4$ kg, $l_c = 4$ m); b — horizontal cylindrical charge ($C = 36$ kg, $l_c = 6$ m).

tances 80–380 r_c (r_c, charge radius). This duration corresponded approximately to the total time taken for the direct and reflected longitudinal waves to pass through distances in the boundary limits of a conditional cone of ejection of soil above the charge. Therefore, if it is to be considered that the formation of a cone of ejection is governed by the essentiality of getting zero or near-to-zero tangential stresses at its boundaries, then the generation process of the above-mentioned low-frequency vibrations could be related to the dome-shaped heaving of the soil.

The characteristic distance 380 r_c is indicated by r_e. In blasting a vertical cylindrical charge in loam, the time of dome upheaval, t_K, is

$$t_K = 0.075\sqrt[6]{CH^2},$$

where H is the depth of the upper end of the charge from the surface, m.

The experiments involved recording the parameters of seismoblast waves at various depths of charges weighing 4 kg. They confirmed the view that a blast is a two-stage source of vibrations: at the surface ($H = 0$ m), at depth ($H = 24$ m), obviously providing the camouflet effect of blasting a given charge, and also at an intermediate depth ($H = 8$ m).

It can be seen from the wave diagrams of these blasts (Fig. 19) that low-frequency vibrations are absent in the oscillograms of first arrivals of the P-wave but in subsequent vibrations beyond the characteristic distance r_e are absent only in the records of surface vibrations induced by blasting a charge placed at depths $H = 0$ m and $H = 24$ m, i.e., in such conditions of blasting wherein the second phase of the blast effect was not felt.

Thus the effect of the second phase of a blast at its epicentre, when a free surface is available, produces ground movement in the shape of a dome

60

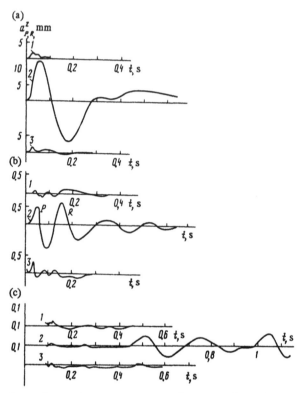

Fig. 19: Dependence of ground movement $a_{P,R}$ on time t, at epicentral distances r (a—3 m, b—15 m, c—150 m) when a charge weighing 4 kg was blasted at different depths H (1—0 m, 2—8 m, 3—24 m).

type upheaval, whose time of heaving is dependent on the quantum and depth of placement of charge. Here ground movements continue for a long period with considerably large amplitudes (Fig. 20), since the soil particles attain insignificant initial velocities and are not retained by elastic forces. Plastic and elastoplastic ground movements as well as the non-linear phenomena associated with them occur in the variation in period of direct longitudinal wave over distance. Initially, in the near-field zone this period decreases, but further in the elastic zone, it remains constant.

A similar picture is seen in the non-uniform dissipation of energy of the longitudinal wave which occurs when the latter passes through near- and far-field zones.

Reduction in the maximum amplitudes of P-wave over distance is usually approximated by a power function of the type $a \approx r^{-n}$. The index of power varies depending on where the wave is propagated: in the near-field zone of blast or in the elastic zone. Thus, for example, when a vertical cylindrical charge is

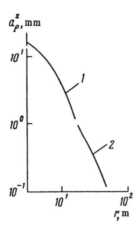

Fig. 20: Dependence of ground movement on distance of P-wave in (1) near-field zone and (2) far-field zone ($C = 4$ kg, $l_c = 4$ m).

blasted (see Fig. 20) in the first section, the displacements attenuate as per the law $a = 160\ r^{-2}$ and in the second section as per $a = 45\ r^{-1.6}$. The radius of zone on whose boundary appears a break of attenuation curve of blast wave parameters, corresponds to the dimensions of the near-field zone of the blast effect denoted by r_e. The completely formed phases of surface wave R are observed partially at such distances.

Based on the results obtained in the investigation of near-field zone of blast effects, it can be emphasised that:
— Directivity of amplitude vector of vibrations towards the blast, i.e., amplitude of vibrations in the direction of the blast, is lower than the amplitude of vibrations away from the blast;
— Attenuation index of amplitudes of vibrations over distance is more in the near-field zone compared to the far-field zone;
— Sinusoidal (elastic) wave is absent;
— Reduction in the period of vibrations in the wave in the near-field zone and its constancy in the far-field zone are observed;
— Dynamic stresses imposed by a longitudinal wavefront of compression up to specific distances r_e exceed the ultimate structural strength of the soil;
— At distances r_e from the blast epicentre, as seen from Fig. 20, a completely formed elastic surface wave exists.

Analysing these peculiarities in the near-field zone of blast effects, it can be said that this zone up to distances r_e can be termed the zone of residual deformation of soil.

Considering the distance r_e at which characteristic peculiarities were observed in the propagation of blast-induced seismic waves, it can be seen that the waves have one and the same values of u (ground movements) or velocities of displacements for particular soil types. Thus, for example, when vertical cylindrical charges were blasted in loam from Kiev, at a distance r_e, u happened to be 3–5 cm/s, while in a blast of horizontal cylindrical charges in loess u was 8–10 cm/s.

Therefore, knowing the structural strength of soil or taking the value of particle velocity of P-wave at the characteristic break point in the curve (at a distance r_e from the epicentre of blast), the dimensions of near-field zone and zone of residual displacement of soil can be calculated from experimental data.

Based on the processing of experimental data in order to calculate the dimensions of the near-field zone on the surface, the following formulae are obtained:

in blasts of vertical cylindrical charges

$$r_e = k_v \sqrt{C_1 H}, \qquad \ldots (3.1)$$

in blasts of horizontal charges

$$r_e = k_h \sqrt{C_1 H_{od}}, \qquad \ldots (3.2)$$

where k_v, k_h are coefficients that take into account the blasting conditions;

C_1 is the linear charge density, kg/m;

H is the depth of charge placement (up to the upper edge), m;

H_{od} is the scaled depth of charge placement, $\text{m/kg}^{\frac{1}{2}}$.

The values of k_v for loams of Kiev are equal to 7–9, in manganese loam 11–13. The values of k_h for Kiev loams 14, for Volgograd loams 12, for Kakhovsk loess 8 and for Kerchenskii clay 26.

For blasting of vertical cylindrical charges, the values of r_e are considered only up to depth H, corresponding to $r_e = 8\sqrt{C_1}$, since with deeper placement of charge the intensity of P-wave remains constant.

The dependence of ground movement (with charges weighing 4 kg) on different depths H is shown in Fig. 21.

In these experiments the zones of irreversible deformation of soils should be equal: with a blast at the surface 9 m; blasts at depth of 8 and 24 m, 15 m. At such distances from the epicentre of the blast, a break is noticed in the curves showing the relationship between variation in displacement amplitudes and distance of blasts at the surface and at a depth of 8 m.

In a blast with the charge at 24-m depth, the zone of irreversible deformations is approximately equal to 15–17 m. As the seismic measuring instruments were installed at the surface, i.e., in the zone of elastic vibrations, the curve shown in Fig. 21 is observed to be continuous.

Larger amplitudes of displacement are noticed in the near-field zone of blast effects, for blasts at 0 and 8 m, vis-à-vis the amplitudes in blasts at 24 m. This is firstly due to the fact that nearer the surface the particles having gained

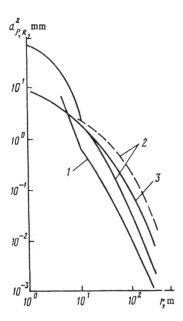

Fig. 21: Dependence of ground movement at surface on distances and depth of charges weighing 4 kg.

$1 - H = 0$ m; $2 - H = 8$ m; $3 - H = 24$ m.

significant initial velocities from the blasts at 0 and 8 m, are not retained in the field of effect of elastic forces and secondly, the amount of energy absorbed due to the propagation of waves through a short distance to the free face is considerably less. In the far-field where the disturbances arrive through lower bedding and highly dense layers, the maximum values of amplitude of soil vibrations are directly proportional to the depth of charge placement.

3.2 Zone of Residual Deformations

A study of the mechanism of the blast kernel, propagation of waves in the near-field zone and trajectory of particle movements demonstrated within the boundaries of the blast centre, plastic and elastoplastic deformations take place, i.e., residual deformations of the massif are observed. The currently available methods of studying the zones of residual deformations help in establishing the dimensions with sufficient accuracy and also certain other quantitative indices. Using the most accurate of all these methods, i.e., the method of radioactive mapping, it was possible to establish that the volumetric deformation of soil is reduced as the distance between the soil and the charge varies according to the power law, which characterises that fraction of energy of products arising out of the blast expended in plastic deformations. Moreover, the absolute zones of residual deformation enlarge as the charge weights increase and, furthermore, also depend on the shape of the charge.

The dimensions of zones for various types of charges in soils of 1.97 T/m^3 density and moisture content 19–20% (loams of Experimental Blast Base of Academy of Sciences, Ukrainian SSR) were determined with the resolution capability of instruments of radioactive mapping (lower limit of measurements up to the 10^{-2} values of volumetric deformations). These zones of concentrated, cylindrical-vertical camouflet and horizontal charges are 50 r_c, 250 r_c and 120 r_c respectively.

Additionally, the dimensions of zones of residual deformation as above and the presence of several structural features in the soil massif do not adequately explain the phenomena associated with the long-term stability of underground and open-cast workings, counterfiltration properties of hydrotechnical structures built by blasting, laws of deformation and propagation of seismoblast waves etc. A study of the laws of propagation of seismic waves within the near-field zone confirmed that the zonal dimensions of residual deformation of soils is several times more than those which can be determined by using standard methods of investigation.

In this connection, an attempt was made to widen the range of measurements that could be carried out by instruments, particularly in respect of the dimensions of the zone deformed by the blast and many other quantitative characteristics of the medium within its boundaries.

Considering the fact that blast waves change their parameters while passing through a heterogeneous soil massif (to which can be attributed also the zones of irreversible deformations induced by blast effects), experiments were conducted under soil conditions as follows:

In an undisturbed soil massif, the parameters of blast-induced seismic waves in test experiments Nos. 7 and 8 were determined (Table 5).

Later, in the same area, a horizontal excavation 6.5 m in length, 2 m in depth and 5.4 m in width at the top was made by blasting a charge (may be dubbed the 'base experiment') (Fig. 22).

Table 5

Nature of blast	Charge parameters				Charge shape
	C, kg	H, m	r_c, m	C_1, kg/m	
Test experiment No. 7	1	10	0.0528	—	Spherical
Test experiment No. 8	0.1	4	0.0245	—	Spherical
Base experiment No. 14	36	1.5	0.0345	6	Cylindrical
Shot experiment No. 15	1	10	0.0528	—	Spherical
Shot experiment No. 16	0.1	4	0.0245	—	Spherical

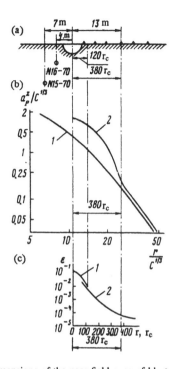

Fig. 22: Determination of dimensions of the near-field zone of blast effects at surface:

a—location of seismic receivers and scheme of placement of charges; b—relationship between variations in ground movement and distances in the near-field zone during (1) test and (2) shot blasts; c—changes in volumetric deformations as distance was varied in the main blast measured by (1) radiometric and (2) seismic methods.

The quantitative parameters of the longitudinal wave P recorded by the monitors VBP-III are reproduced below:

The parameters of longitudinal wave and deformation characteristics of the soil massif in the main blast:

Epicentral distance, m	3	6	9	13	16
Soil displacement a, mm	87	15.2	2.25	0.64	0.6
Mass velocity of displacement u, cm/s	460	48	13	4.6	3.9
Velocity of wave propagation, v_P, m/s	200	350	500	520	540
Volumetric deformation, ε	2×10^{-2}	1.36×10^{-3}	2.6×10^{-4}	8.9×10^{-5}	7.2×10^{-5}
Radial stress, $\sigma_r \cdot 10^{-5}$, Pa		18.2	3.23	1.26	0.461	0.405

Soon after the main blast, without rearranging the seismic recorders, blast Nos. 15 and 16 (dubbed 'shot blasts') were conducted, whose waves passed through the already deformed rock (as a result of the main blast).

Values of soil particle displacements in the wave during test and shot blasts are shown below:

Epicentral distances, corresponding to profile of measurements in the main

blast, m	0	3	6	9	13
Test drill hole Nos. 7, 8	0.775	0.6	0.425	0.29	0.2
Shot drill hole Nos. 15, 16	1.44	1.29	0.975	0.63	0.23

Ground movement in shot blasts within the distance range of (0–400) r_c were 1.8–2.3 times more than in test blasts at the same distances. This signifies the existence of residual deformations in the rocks in that particular region induced by the main blast. Here the soil particle displacements for longitudinal waves of shot blasts decrease as the distance from the main charge in the direction of the elastic zone increases, i.e., such a process is observed which is qualitatively similar to the one established by standard methods of investigation into the squeezed zone of soil subjected to blast effects.

With such an assumption regarding the zone of residual ground deformation, the parameters of the direct longitudinal wave may be stipulated and the deformation characteristics of the massif quantitatively determined by using the relationships:

$$\varepsilon = u/v_P;$$

$$\sigma_r = uv_P\rho,$$

where σ_r is the radial stresses at the wavefront;

$\rho = \rho_0/(1 - \varepsilon)$ is the instant density of rock at the wavefront;

ρ_0 is the initial rock density.

To explain the deformational characteristics of the soil massif during the main blast, a radiometric instrument (GGP-1) was used in addition to seismic instruments. Moreover, the seismometric profile was a continuation of the radiometric profile in the direction of lower deformations.

It is possible to evaluate soil stresses in the near-field zone of blasts using the quantitative data on blast wave parameters and the above-mentioned relationships. Thus at a distance 380 r_c from the epicentre of the main blast the radial stresses at the longitudinal wavefront can be computed as $\sigma_r = 0.46 \times 10^5$ Pa, while the ultimate triaxial compressive strength of Kiev loam was found to be $\sigma_S = (0.4 \text{ to } 0.6) \times 10^5$ Pa. In other words, the distance approximately equal to 400 r_c happens to be the boundary of the zone of inelastic vibrations of soil for the given blast.

The seismometric method developed for determining the zone of irreversible deformations in soils enables one to distinguish clearly the changes in permeability, strength and other physicomechanical properties of soil. This method was used in establishing the changes in the aforementioned properties of the soil massif around a main canal constructed by means of blasting. In a particular section of the canal, where measurements were taken, the linear charge density used in its construction was 80–90 kg/m and the depth of placement was 2 m.

Profiles of transducers used to measure the seismic wave parameters in a specific zone of residual deformation of soils, were laid along the canal axis — 4 profiles spaced 13, 28, 40, 50 m respectively from the canal axis and one also across it (Fig. 23).

As can be seen from Fig. 24, at distances of 13–50 m (85–330 r_c), the soil displacement in shot blasts was 1.2–1.7 times more than in control blasts at the same distances, decreasing later in a direction away from the canal axis.

The dimensions of zone of residual deformations in soils, as measured by the seismometric method, correlated well with the following data: on the variation in physicomechanical properties, obtained through the method of electrical prospecting (using induced polarisation station VP-62); on the changes in filtration properties of soils (using Znamenskii's apparatus); and on the changes in density-related properties of soils obtained by radiometric mapping.

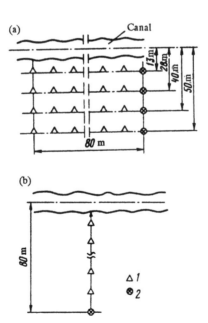

Fig. 23: Diagram for determining dimensions of near-field zone of blast effect, installation of seismic recorders and placement of shot charges, when measurements were taken along the profiles: a — along the canal axis; b — across the canal axis; 1 — seismic transducers; 2 — shot charges.

68

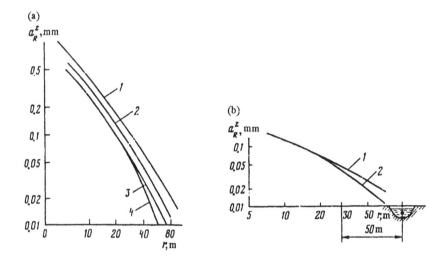

Fig. 24: Relationships between soil particle displacements in the surface wave a_R^z and distances up to the blast site r:

a — in shot blasts along the longitudinal profiles of measurements: 1 — 13 m from the canal axis; 2 — same, at 28 m; 3 — same, at 40 m; 4 — same, at 50 m; b — (1) in shot blasts and (2) in test blasts along the transverse profiles of measurements.

The data obtained enabled us to search more judiciously for solutions to the problem of long-term stability of underground and open-cast engineering structures in squeezed rocks without supports and to calculate the reliability of a counter-filtration screen of compacted soil as a result of blasts.

Such investigations help additionally in in-depth studies of the nature of individual waves, mechanism of excitation and accurate calculation of the seismic effects of a blast.

3.3 Pattern of Propagation of Cylindrical Blast Waves within a Soil Massif

Experimental investigations into the parameters of blast waves in the near-field zone, at the free face and within the massif, are of particular interest in the following aspects:

(i) in the study of phenomena related to the volumetric deformability of soils under the blast effect of cylindrical vertical charges and

(ii) formation of seismic waves.

A series of experiments were conducted in the Experimental Blasting Base of the Academy of Sciences, Ukrainian SSR, to investigate the patterns of ground movements within the medium and to study the characteristic peculiarities in the variation of soil particle vibration parameters with increasing depth of charge placement and distances from the axis of cylindrical vertical charge. This was done at a depth H using VBP-III and VIB-A instruments.

The VBP-III was oriented towards the blast source and for recording the vertical and radial components in ground movements. It was placed in drill holes of 350 mm diameter, at 0–8 m depth from the surface, and at distances of 3–28 m from the axis of the charge (Fig. 25), which amounted to distances of 210 r_c and 2000 r_c.

In the experiments, vertical cylindrical charges of 4-m length having a linear density of 1 kg/m were blasted. The charges were initiated from both ends simultaneously. Compacted trotyl was used in these test blasts.

The oscillogram of grouped movements at a distance of 210 r_c (Fig. 26,a) and trajectories of soil particles at distances of 210 r_c and 2000 r_c (Fig. 26,b) from the axis of the cylindrical charge, are shown in Fig. 26. From the figure a sharp difference is observable in the form and magnitude of ground movements along the vertical component at the free face and within the medium.

The amplitude of displacements at the top decreases continuously with an increase in depth. This is visually confirmed by Fig. 27, in which are shown the variation in ratios of amplitudes of soil particle displacements in the direct wave at a current depth H and at the surface along the z- and x-components.

The displacement amplitudes along the horizontal component at the same distances vary in an entirely different manner with an increase in depth. With depth, the amplitudes gradually increase, especially up to the conditional centre of gravity of the charge. This can be explained by the superposition of low-frequency vibrations on the process induced by a compressive wave in this zone, as a consequence of the effect of the second phase of the blast (back pressure of detonation products). Moreover, the oscillograms of blast and the trajectories of soil particle movement (Fig. 26) show that the displacements in the direction away from the blast are more than the displacement towards it. This confirms the inelastic nature of deformations in the massif in the area under consideration. Further increase in depth results in a gradual reduction in soil displacements to values which are many times lower than those observed at the free surface (at equal distances relative to the ends of vertical charge).

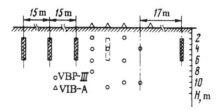

Fig. 25: Sketch showing location of seismic receivers in the investigation of blast wave propagation within a soil massif.

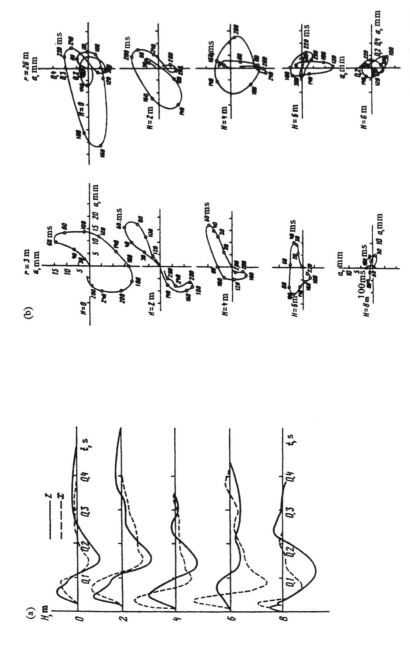

Fig. 26: (a) Displacement and (b) trajectories of particle movement at the free face and within the soil massif when a vertical, cylindrical charge is blasted. ($C = 4$ kg; $l_c = 4$ m).

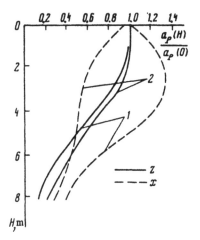

Fig. 27: Variation in ratios of displacement amplitudes $\dfrac{a_P(H)}{a_P(O)}$ in the body wave at a current depth H and at the surface along z- and x-components ($C = 4$ kg; $l_c = 4$ m): 1 — at a distance of 210 r_c; 2 — at a distance of 2000 r_c.

Thus at distances nearer to the blast source (barring the area within the soil massif which is equidistant from the cylindrical vertical charge), the same law applies to the qualitative variation in soil displacements both in the vertical and horizontal plane, and in particular to the intensity of displacements, which decreases with depth. At these distances the axial symmetry of field of deformations is distinctly observable. This is associated with the axial symmetry of the type of charge being investigated.

This action of a cylindrical vertical charge on the soil massif is clearly seen in the curves showing the variation in velocities of soil particles with depth. A distinct pattern emerges here which is associated with increase in the intensity of vibrations from the level of the free face to the centre of gravity of the cylindrical vertical charge and subsequent gradual reduction in intensity. The maximum displacement velocity is observed at the central zone of charge symmetry.

The curves showing changes in ratio of values of mass velocities in the P-wave within the near zone at a depth H, for an analogous index at the free surface as the depth is increased are approximated by the following relationships:

along the z-component $k_z = 0.83\ H^{2.75} e^{-0.85H}$ and

along the x-component $k_x = 1.03\ H^{2.85} e^{-0.71H}$.

The experimental data, approximated by the above relationships, have considerable practical significance. For example, in solving a series of engineering problems related to the forming of vertical workings in compacted soils by means of blast energy, the blasting operations are often conducted in the immediate vicinity of various civilian and industrial structures. This is associated with considerable loads being exerted on building foundations. Based on the above data, the location of the upper end of the cylindrical charge can be chosen such that the building foundations will be within the effective zone of minimal possible blasting loads for a particular charge. Furthermore, the practical utility of the relationships lies in the fact that by using them the intensity of expected dynamic loads within the massif can be calculated if we have similar data pertaining to analogous geometrical points at the surface level (obtaining data at the surface level through experiments is far easier). This aspect has much significance in calculating the stability of the various underground structures built by blasting operations and also in developing technological procedures based on using the effect of the 'middle zone'.

As the distance from the blast site increases, the axial symmetry of the effect of vertical cylindrical charge on the soil massif becomes less and less and at certain distances such a charge forms a field of velocities analogous to a charge with central symmetry. Elastic vibrations are observed at distances of $2000 \, r_c$ and more. They are indicated by the elliptical type movement of soil particles with absolute values along the z-component dominating over those of the x-component (see Fig. 28,b) as well as by the presence of a new type of oscillatory movement, a low frequency wave of the Rayleigh type (R). The latter aspect is characterised by the shift in direction of movement of soil particles: clockwise movement turns to anticlockwise at time $t = 0.14$ s compared to the initial stage of the process.

The experimental data also confirm the currently held views on twin effects of a blast located near the free surface. From soil particle trajectories and other time-related parameters of the process it is evident that the process of vibrations excited by such a charge is governed by the effect of charge in the role of a source of the expansion type and further is associated with dome-shaped movement of soil at the surface. The specific characteristic indicated above is evidenced more distinctly near the charge and at the surface level. With increase in depth, the influence of dome-shaped heaving of the soil on the process of vibratory movement reduces. This is partly confirmed by the fact that with an increase in depth the duration of ground movement in a direction away from the blast decreases. However, the influence of a free face and the related dome-shaped heaving of soil above the blast kernel, bear upon the processes of vibratory movement even at the level of the conditional centre of gravity of a cylindrical vertical charge. This can be seen from the trajectories of soil particle movement. Such a charge acts on other areas more like a source of the expansion type.

4

Seismic Effects due to Blasting
of a Cylindrical Charge

4.1 Methodology of Investigations

Blasting technology using extended cylindrical charges has found widespread application in open-cast mines as well as in hydro-reclamation installations. In such operations, the task of conducting blasts from a seismic safety point of view is of paramount importance, especially if the blasts are carried out for reconstructing the operating units or in the vicinity of residential units, power lines, gas and petroleum pipelines etc. Further, the seismic safety of blasting operations has become a topical theme for study since no method is currently available for calculating the seismically safe distances when such charges are blasted in compacted soils.

Certain patterns in the propagation of cylindrical blast waves in compacted soils having a free face were determined based on the research work undertaken at the S.I. Subbotin's Institute of Geophysics (Academy of Sciences, Ukrainian SSR). Relationships were obtained for determining the parameters of blast-induced seismic waves and seismically safe methods of conducting blasting operations were evolved.

Standard seismometers were used to record the ground vibrations: seismic receivers VBP-III, VIB-A, VEGIK and magnetoelectric oscillographs H-700. The magnitude of displacements and rate of displacement were the main parameters recorded.

The magnification curve of measurements was recorded while calibrating the apparatus on a vibrating table as well as by calculations. Moreover, to check the accuracy of readings given by instruments, periodic blasts were conducted and the reproducibility of results checked.

Two measurement profiles were scanned in the process of investigations: in the direction perpendicular to the axis of the charge and along the axis of the charge. At each observation point two components of vibrations were recorded: vertical z and horizontal x. Seismic receivers were placed at intervals of $\Delta \tau_{n+1} = (1.2 - 1.5)\Delta \tau_n$. Ground vibrations were observed both during experimental and industrial blasts of vertical and horizontal charges (with a free face). The epicentral distances for installing seismic receivers were dependent on the charge mass and covered a range of distances from 0 to 4000 m. The charge mass varied from 2 kg to 400 kg and its length from 2 to 1300 m. Data concerning constructional features of the vertical cylindrical charges are given in Table 6, while data for the horizontal cylindrical charges, concentrated charges and soil characteristics of experimental and field blasting are given in Table 7.

In studying the effect of blasting cylindrical charges, scaled and deep-seated blasts were conducted. The former was carried out by placing charges of different weights at the same scaled depth and the latter with charges of the same weight placed at different depths.

In the experimental investigations the distance from the surface level up to the centre (centre of gravity) of the charge was assumed as the depth of placement of the vertical cylindrical charge. This assumption is justified within the limits of accuracy of measurements related to the far-field (elastic) zone of blast effects.

The parameters of longitudinal P- and surface R-waves were studied separately, depending on mass C, radius r_c, length of charge l_c, depth of its placement H, epicentral distance r and conditions of blasting.

The longitudinal wave is the dominant component according to intensity of vibrations in the near-field zone of blast effects and the surface wave in the far-field elastic zone. By near-field zone is meant the zone of residual deformations of soils, whose dimensions can be determined by relationships (3.1) and (3.2).

The coefficients k_1 and n were obtained as by processing the experimental data with the coefficient of variation 6–19%. This confirms the accuracy of measurements pertaining to the vibration parameters of rocks during blasts.

Table 6

S. No. of blast	C, kg	l_c, m	r_c, m	l_{bh}, m
1	585	37	0.055	42
2	350	24	0.055	31.5
3	180	12	0.055	18
4	90	6.5	0.055	12.5
5	27	7	0.028	10
6	16	4	0.028	6
7	12	4	0.024	6
8	12	6	0.02	10
9	12	6	0.02	8
10	12	6	0.02	6
11	12	12	0.014	14
12	8	8	0.014	12
13	8	8	0.014	10
14	4	4	0.014	6
15	4	4	0.014	5.5
16	2	4	0.01	6

Table 7*

S. No. of experimental site	Soil characteristics				Parameters of charge			Values of coefficients	
	Type	Density $\rho \cdot 10^{-3}$, kg/m³	Moisture by weight, %	Porosity, %	Charge weight, C, kg	Scaled r_0, m/kg$^{1/3}$	Scaled depth H_0, m/kg$^{1/3}$	k_1	n
1	Loess (test operations in the construction region of Kakhovsky irrigation system)	1.55	7.2	43	160–800	2–21	0.32–0.59	25	1.4
2	Loam, dry sand (in the region of construction of the collector 'central' Astrakhan)	1.69	8	34	2.5×10^4	13.5–83.5	0.46–0.5	75	1.3
3	Light loam, loamy sands, sand (in the construction area of Toguskenskii canal, Kazakhstan)	1.75	9	38	2.33×10^5 to $\dfrac{4 \times 10^5}{6}$	$\dfrac{0.55–81.7}{6–66}$	$\dfrac{0.55–0.63}{0.275}$	85	1.3
4	Heavy loam (section in the construction of 'Volgaural' canal)	1.93	14	36	1.75×10^5 to 2.46×10^5	20–50	0.27–0.35	100	1.3
5	Dusty loam (Academy of Sciences Polygon-Ukraine Kiev region)	1.67	14.2	35.2	$\dfrac{8–36}{0.1–12}$	$\dfrac{15–145}{5–200}$	$\dfrac{0.4–1}{0.55–15}$	90	1.3
6	Heavy loam, clay (in construction area of Palasovskii canal)	1.74	10.8	34.8	$\dfrac{6; 12}{6; 12}$	$\dfrac{9–75}{6.75}$	$\dfrac{0.4–0.5}{0.22–0.275}$	125	1.3
7	Water-saturated clayey soil (in construction region of drainage canals in Kubansk)	1.85	36	51	$42; 3 \times 10^3$	3–86	0.16–0.38	120	1.05

* Reproduced as given in Russian original, although some figures appear incorrect — General Editor.

4.2 Relationship between Wave Parameters and Charge Shape

In order to understand the characteristic features (qualitative and quantitative) of seismic effects due to the blasting of a cylindrical charge, comparative studies were conducted with the already established seismic effects due to the blasting of a spherical charge.

The qualitative aspects of the vibration process in soils due to blasts are shown by the wave diagrams in Fig. 28.

An analysis of the experimental data showed that from a qualitative viewpoint, the wave processes in blasts of cylindrical charges are similar to the processes observed in seismic investigations of blasts with spherical charges.

Initially such a wave was fixed whose time-dependent and absolute parameters were to be compared with the results of parallel measurements by velocity transducers of the electromagnetic type. These transducers are normally used to record shock waves. Such a wave fixation enabled us to identify it with the longitudinal body wave (phase P). This wave attenuates over distance and with time the maximal amplitudes of vibrations were observed at characteristic distances (boundary of the zone of elastic vibrations), at different phases of the low-frequency wave, which attenuated rapidly with depth. The wave dampened over distance as per the law $a \approx r^{-1.2}$. The dominant periods were 0.15 –0.25 s (for the studied range of charge weights) and the velocity of propagation in Kiev loam was 350–400 m/s.

The characteristics described above made it possible to refer to the low-frequency wave as a surface wave of the Rayleigh type. The individual phases R_1, R_2 appeared in the oscillograms. In the active zone of surface wave R the process of 'pumping' of energy into the subsequent phases is observable with an increase in distance, i.e., gradual attenuation in amplitudes at initial phases and its propagation in subsequent phases.

As can be seen from Fig. 28,a, the amplitude of vibrations along the z-component in the P-wave in the near zone exceeds that of the x-component at the same measurement points. This is explained by the directional effect of the second stage in the blast towards the free face.

In blasts of spherical charge (Fig. 29,a) and cylindrical charge (Fig. 29,b), at close distances the particles begin to move along a curve rotating clockwise and at far distances the movement follows an elliptical orbit rotating anticlockwise. The time periods $t = 35$ ms in a blast of cylindrical charge and $t = 55$ ms in a spherical charge, in the particle trajectories coincide with the time of approach of the positive phase of compression wave to the free face and the return of reflected wave to the walls of the cavity when the massif begins to be unloaded and begins to move towards the free face. The time periods corresponding to the maximal values of stress wave of compression at similar distances in analogous soil conditions, coincide with the values indicated in the soil particle trajectories, with a particular time lag in the deformations due to stresses in the compression wave. These periods are measured by strain-gauge pressure sensors.

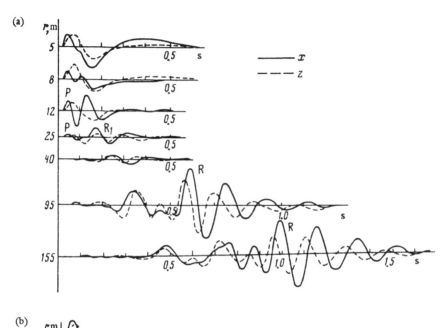

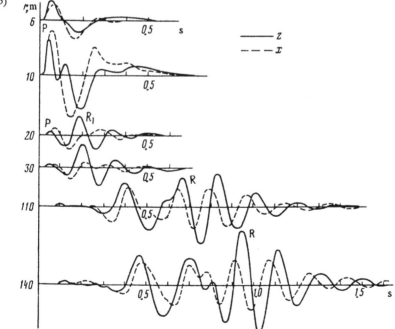

Fig. 28: Wave diagrams obtained in blasting of (a) cylindrical and (b) spherical charges ($C = 4$ kg; $l_c = 4$ m).

Subsequently, the movement induced by the effect of the blast products becomes superposed on the aforesaid ground movement over time $t = 60$ ms (cylindrical charge) and $t = 75$ ms (spherical). This evident superposition is directed towards the blast. The effect of superposition decreases as the distance between observation points on the free surface and the source of disturbance increases.

The inelastic nature of this vibration is clearly seen in the oscillograms and soil particle trajectories: displacements in the direction of the blast are less than those away from the blast.

Attention is drawn to the fact that the period of rarefied wave marked in the oscillograms of blasts, is determined by the range of time fixed in the soil particle trajectories: 75–300 ms and 50–260 ms for spherical and 60–250 ms and 50–280 ms for cylindrical charges at corresponding epicentral distances of 5 and 10 m. This duration corresponds to the period during which the soil while moving in the epicentral zone still maintains links with the surrounding massif.

Further, the low frequency vibrations recorded in the near-field zone of blast characterise the inelastic deformations of soil in that zone.

At distances of the order 50 m from the blast epicentre, i.e., in the zone of

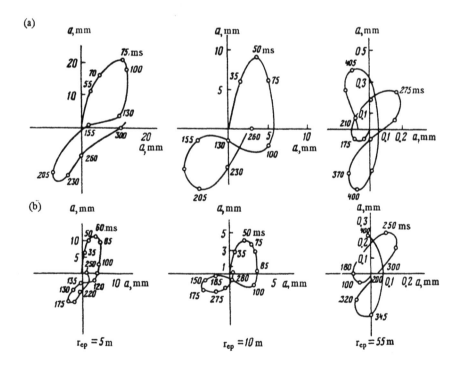

Fig. 29: Soil particle trajectories in blasts of (a) spherical charges and (b) cylindrical charges for various epicentral distances $r_{ep}(C = 4$ kg; $l_c = 4$ m).

elastic vibrations, soil particles move in elliptical trajectories, both in the case of spherical and cylindrical charges (Fig. 29). This is obviously due to a shift in phase of horizontal component relative to the vertical by a quarter period. Initially the movement is directed clockwise but at a particular time period the soil particles begin moving in an anticlockwise direction. A detailed study of the trajectories of ground in all types of blasts and also wave diagrams showed that the change in direction of particle movement occurred at the period a new phase in wave movement had developed — in the given case, the new phase of surface wave.

Typical wave diagrams pertaining to blasts of elongated charges are shown in Fig. 30. The direct longitudinal P-wave is noticed in the early arrivals of the wavefront. The x-component of this wave initially exceeds that of the z-component. In the far elastic zone the absolute values of vibration amplitudes equalise initially and later the z-component exceeds the x-component. The P-wave attenuates notably over distance and time and from specific distances r_e, maximum amplitudes of vibrations are observed in various phases of the R-wave.

It is to be noted that in investigations of seismic effects induced by blasting elongated charges at a site with a stratified geological structure, low-frequency reflected waves P_{Lref} comparable in intensity with surface long-period waves were observed in the seismogram at the indicated distances (Fig. 30,b). The fact that these were not surface waves is confirmed by their higher velocity of propagation compared to those of reflected waves. Soil particle movement was directed clockwise (Fig. 31), while in the surface wave, soil particles moved in the anticlockwise direction.

These comparisons between spherical and cylindrical charges according to wave diagrams and soil particle trajectories refer to the qualitative characteristic of blasts. It is evident that blasts of spherical and cylindrical charges are principally the same in their qualitative indexes.

Distinct differences were also noticed quantitatively in the wave parameters related to blasts of cylindrical and spherical charges for the same set of conditions. For example, in the near zone of blast effects, the intensity of vibrations of the longitudinal wave in a blast of cylindrical charge was 2–2.5 times less than in a blast of spherical charge of the same weight under similar conditions. This is confirmed by Table 8 in which data on ground movements induced by blasts of spherical and cylindrical charges are given. The intensity difference in longitudinal waves gradually reduced and the trend was maintained up to distances equal to six times the length of the cylindrical charge; subsequently, the intensity of vibrations of cylindrical and spherical waves equalised (Fig. 32).

While comparing the quantitative indices of surface waves, similar peculiarities were noticed in the nature of wave propagation.

(a)

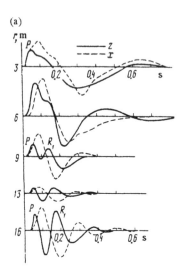

(b)

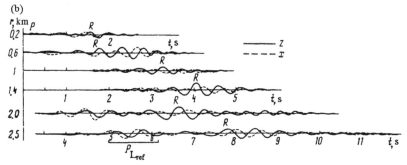

Fig. 30: Wave forms pertaining to blasts of elongated horizontal charges:
A—near-field zone ($C = 36$ kg, $l_c = 6$ m); b—far-field zone ($C = 2.5 \times 10^4$ kg³; $l_c = 300$ m).

Thus, if one were to consider that the region of equal values of wave parameters, such as displacement or rate of displacement of particles in a blast of two charges having the same characteristics with axial and central symmetry, is identically equal to the region of transition of cylindrical wave into spherical, then from the present investigations one could draw scientific and practical conclusions about the transition of a cylindrical wave into a spherical at an epicentral distance approximately equal to 6 l_c.

The shifting of epicentre of blast from the transition region of a cylindrical wave into a spherical depends on the form of cylindrical charge, i.e., on the ratio of charge length to its diameter and depth of emplacement. In the present studies, charges having an indicated ratio of not less than 125 were considered.

We can now examine the characteristic features of blast effects of cylindrical

Table 8

	Cylindrical charge			Spherical charge	
r, m	a, mm		r, m	a, mm	
	P	R		P	R
6	4	—	6	8.3	—
8	2.4	—	8	4.6	—
10	1.5	—	10	4	--
12	1	—	12	4.2	—
16	0.38	0.8	16	1.4	1.12
20	—	0.59	20	0.32	0.91
30	0.09	0.5	30	0.14	0.63
40	0.025	0.35	40	0.04	0.4
63	—	0.25	63	—	0.25

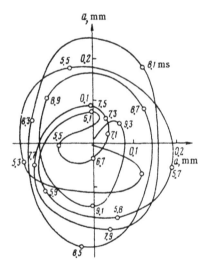

Fig. 31: Soil particle trajectories in a blast of horizontal charge at a distance of 2500 m from the blast site ($C = 30$ T; $l_c = 300$ m).

charges in reference to the shape of the charge and its placement relative to surface.

In the near-field zone of blast effect induced by a cylindrical charge, the amplitude of vibrations depends largely on the charge radius (in other words, on the linear mass of charge) and to a lesser extent on the total mass of charge. For example, the relationships between variation in displacement velocities and distances in blasts, of two horizontal cylindrical charges with the same linear charge mass of 20 kg/m conducted in loess are shown in Fig. 33. The total mass of charges was 680 and 200 kg and lengths 34 and 10 m respectively.

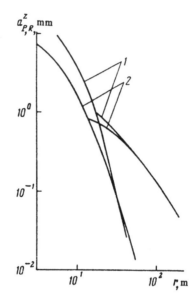

Fig. 32: Variation in ground movement $a^z_{P,R}$ with distance r in blasts of (1) spherical charge and (2) vertical cylindrical charge ($C = 4$ kg).

Despite the difference in charge weights, the intensity of vibrations in blasts of these two charges was found to be same in the near-field zone. In the far-field elastic zone, where the entire weight of the charge plays a significant role on the seismic effects of the blast, the intensity of vibrations in a blast of charge weighing 600 kg is considerably more than that of a charge of 200 kg.

The asymmetry in intensity of vibrations in the direction along the charge axis and perpendicular to it is also a characteristic feature of the blast effect of a cylindrical charge. As can be seen from Fig. 33 and Table 9, the intensity of seismic vibrations in the surface wave along the axis of charge is 2–3 times less than the intensity at the same distances in a direction perpendicular to the axis of charge. In industrial blasts, asymmetry in the intensity of vibrations along the indicated directions increases by 7–8 times.

4.3 Dynamic Characteristics of Waves in Near- and Far-Field Zones of a Blast

Vertical Cylindrical Charges: Studies of the variation in parameters of waves generated by blasting a cylindrical charge having a free face showed that these parameters depend on the constructional features of charge, its mass, depth of placement, orientation relative to free surface, blasting conditions and the distance up to the observation point. This is confirmed by experimental data on blasts of cylindrical charges given in the previous sections and also by the relationships between variation in ground movement over distance, obtained in

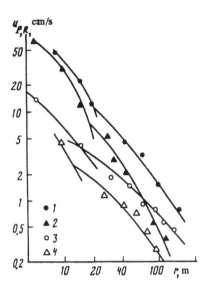

Fig. 33: Variation in displacement velocities $u_{P,R}$ with distance r in blasts of horizontal cylindrical charges along the profile of measurements, perpendicular to the axis of charges and along their own axis (3 and 4) respectively ($1 - C = 680$ kg; $l_c = 34$ m; $2 - C = 200$ kg; $l_c = 10$ m).

Table 9

Weight of charge, kg	Length of charge, m	Linear weight of charge, kg/m	Distance, m	Velocity of ground movement (cm/s) in the directions	
				Along axis of charge	Perpendicular to axis of charge
680	34	20	8	8.7	46.1
			15	4.41	23.15
			40	1.78	5.4
			80	0.87	2.14
			125	0.5	1
200	10	20	8	5.2	39.6
			15	2.4	14.51
			40	0.89	2.7
			80	0.35	0.74
			125	0.16	0.24

blasts of vertical cylindrical charges of scaled series with scaled depth of charge placement $H_{oc} = 2.2$ m/kg$^{1/3}$ = const and charge weights $C = 4$ to 585 kg (Fig. 34,a), and depth of series $H_{oc} = 1.32$ to 3.05 m/kg$^{1/3}$ and $C = 12$ kg (Fig. 35).

84

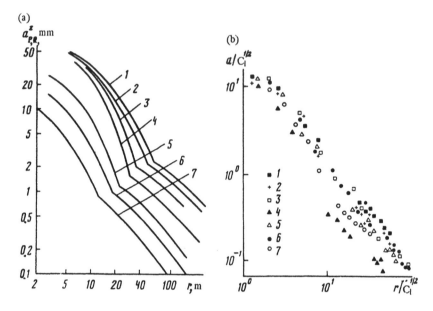

Fig. 34: (a) Variation in ground movement and (b) variation in maximal displacement reduced to $C^{\frac{1}{2}}$ over distance in blasts of mass series of charges:

$1-C = 585$ kg; $l_c = 37$ m; $2-C = 350$ kg; $l_c = 25$ m; $3-C = 180$ kg; $l_c = 12$ m; $4-C = 90$ kg; $l_c = 7$ m; $5-C = 28$ kg; $l_c = 7$ m; $6-C = 12$ kg; $l_c = 6$ m; $7-C = 4$ kg; $l_c = 4$ m.

Analysis of the experimental data enabled us to assume the charge radius as a similarity criterion in the near-field zone (or, equivalently, linear mass of charge). Furthermore, it is well known that clustering of experimental points around a curve related to specific blasting conditions, is an indicator of similarity in the charges blasted.

The relationships between the variation in ground movement with distance, (shown in Fig. 34,b) are scaled to the linear mass of charge in blasts of vertical cylindrical charges in soils possessing different properties. The close clustering of points around the curves signifies the similarity criteria manifesting itself in the linear mass of charge in the near-field zone for blasts of various cylindrical charges. It may be noted that, the scaling of functions of type $a = f(r)$ in the near-field zone of blast effects induced by cylindrical charges to the mass of charges in the form $C^{1/3}$, which is adopted in analysing the blast effect of a cylindrical charge, is not applicable in the given case. It may be applied only to the far-field elastic zone of blast effects of a cylindrical charge.

After mathematical processing, experimental relationships of the type $a = f(r)$ for a longitudinal wave may be approximated by the following formulae, with reasonable accuracy for engineering calculations:

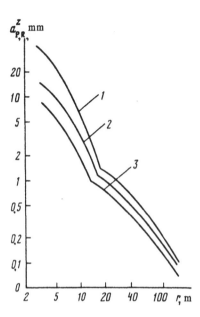

Fig. 35: Relationship between parameters of blast-induced seismic waves and distance and depth of placing charges at constant values of l_c, C and r_c:

$1-C = 12$ kg; $H_{oc} = 3.05$ m/kg$^{1/3}$; $2-C = 12$ kg; $H_{oc} = 2.2$ m/kg$^{1/3}$; $3-C = 12$ kg; $H_{oc} = 1.32$ m/kg$^{1/3}$.

for vertical cylindrical charges in the near-field zone

$$a_P^z = kH^{0.9}C_1^{\frac{1}{2}} \left(r/C_1^{\frac{1}{2}} \right)^{-1.31} ; \qquad \dots (4.1)$$

in the far-field zone

$$a_P^z = 9H_{oc}^{0.9} \left(r/C^{\frac{1}{2}} \right)^{-1.6} , \qquad \dots (4.2)$$

where k is the coefficient for the given set of blasting conditions (soil, Kiev loam; explosive, trotyl), equal to 13;

H is the depth to the upper end of the charge measured from the surface, m;

C_1 is the mass of linear charge, kg.

At distances $r = 6\,l_c$, dampening of maximum displacements with distance occurs according to the laws related to a spherical charge.

The time-dependent characteristic of a longitudinal wave, which describes its propagation more comprehensively, is the time of growth of displacements to the maximum τ_t (Fig. 36), which gradually reduces as the wave propagates from the epicentre to the boundaries of the near-field zone and to those of the zone of residual deformation of soils. Within the elastic zone, this time of growth τ_t depends not on distances, but only on the properties of rocks and charge weights and is expressed by the function

$$\tau_t = 0.014 \; C^{0.13}. \qquad \qquad \dots (4.3)$$

In the elastic zone of ground vibrations, the major portion of the seismic energy is carried by surface waves, characterised by longer periods and amplitudes of vibrations compared to the body waves recorded in the same zone.

Attenuation of ground movement in the R-wave over distances for different constructional features and the extent of penetration of edge of the vertical cylindrical charge into compacted soils of the Kiev loam type, is expressed by the relationship

$$a_P^z = 5.5 H_{oc}^{0.54} \left(r/C^{1/2} \right)^{-1.18}. \qquad \qquad \dots (4.4)$$

Starting from distances $r = 6 \; l_c$, wherein the cylindrical wave transforms into a spherical one, the maximum displacements in the surface wave attenuate over distance according to the law related to a charge with central symmetry.

As can be seen from Fig. 37 and the expression

$$T_R = 0.16 \; C^{0.06} \; r^{0.03}, \qquad \qquad \dots (4.5)$$

the period of vibrations in the surface wave in blasts of vertical cylindrical charges depends on ground conditions, charge weight and distances.

Horizontal Elongated Charges: As already mentioned, the technology of making open excavations, of constructing irrigation and hydro-reclamation installations by blasting of horizontal cylindrical charges, has distinct advantages over the technology that uses concentrated charges. Yet methods for calculation

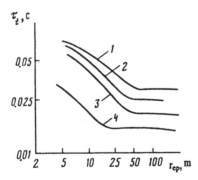

Fig. 36: Changes in time of growth of displacements to maximum value in a longitudinal wave, depending on the charge weight and epicentral distance r_{ep}:

$1 - C = 350$ kg; $l_c = 25$ m; $2 - C = 90$ kg; $l_c = 7$ m; $3 - C = 16$ kg; $l_c = 4$ m; $4 - C = 2$ kg; $l_c = 4$ m.

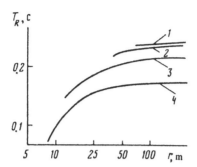

Fig. 37: Dependence of the period of ground vibrations T_R in surface wave on distances and charge weight in blasts of vertical cylindrical charges in loams:

$1 - C = 350$ kg; $l_c = 25$ m; $2 - C = 90$ kg; $l_c = 7$ m; $1 - C = 16$ kg; $l_c = 4$ m; $4 - C = 2$ kg; $l_c = 4$ m.

of seismic-safe distances in blasts of horizontal extended charges are almost non-existent. In the few papers published on this topic [21, 34, 35] no generalised formulae are given. Based on modelling of seismic waves using specific similarity criteria, it appears possible to describe the patterns of ground vibrations in blasts of various elongated charges within the limits of the zone of transition of a cylindrical wave into a spherical, and that too in soils of various physicomechanical properties.

The seismic effects of blasting a spherical charge have been fairly well studied and models of wave parameters developed, especially for the rate of soil particle displacement according to the similarity criterion $C^{1/3}$.

Up to the epicentral distances equal to $6\ l_c$ the seismic effects of blasting a horizontal elongated charge are determined only by the energy of the specified 'effective' section of the charge, i.e., by an equivalent mass of a conditional concentrated charge (a computed quantity) inducing an effect on a point being considered in the massif, which is similar to the effect produced by a real horizontal elongated charge. A relationship was obtained to determine the equivalent mass of a horizontal elongated charge that would act on the points under study in a massif in a direction perpendicular to the axis of charge:

$$C_h = 0.43\ C_{t.h} e^{0.15r/l_c}, \qquad \ldots (4.6)$$

where $C_{t.h}$ is the total mass of a horizontal elongated charge.

The relationships between velocities of ground movement and the scaled distance in blasts of horizontal elongated charges at various experimental sites are shown in Fig. 38 (See Table 7). The quantity $r/C_h^{1/3}$, m/kg$^{1/3}$ is the scaled distance up to the epicentral distances equal to $6\ l_c$ and after these distances the scaled distance is $r/C^{1/3}$, m/kg$^{1/3}$.

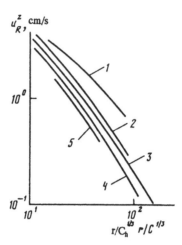

Fig. 38: Relationships between velocities of ground movement and scaled distance in blasts of horizontal extended charges at various test sites:

1—No. 7, $C = 2.68 \times 10^3$ kg; $l = 30$ m; $C = 2.95 \times 10^3$ kg; $l = 130$ m; 2—No. 6, $C = 6$ and 12 kg; $l = 6$ m; 3—No. 5, $C = 8$ kg; $l = 8$ m; 4—No. 2, $C = 2.5 \times 10^4$ kg; $l = 300$ m; 5—No. 4, $C = 1.75 \times 10^5$ and $C = 2.4 \times 10^5$ kg; $l = 300$ m.

As is evident from Fig. 38, the experimental points cluster closely around the corresponding curves, which implies that the similarity criteria $(C_h^{1/3})$ for horizontal extended charges up to the region of transition of a cylindrical wave into a spherical are valid.

Guided by the energy law of similarity as propounded by Acad. M.A. Sadovskii and by our investigations in the field of blasting cylindrical charges, we essayed a formula to determine the ground movement velocity (intensity of seismic vibrations) in a cast blast of horizontal elongated charge in a direction perpendicular to the axis of charge

$$u_R^z = k_1 H_{od}^{0.3} (r/C_h^{1.3})^n, \qquad \dots (4.7)$$

where k_1 and n are coefficients that take into account the properties of the soil to be blasted and the damping of vibrations over distance (to be determined from Table 7) respectively;

H_{od} is the scaled depth of placement of horizontal elongated charge ($H_{od} = H/C_l^{1/3}$, m/kg$^{1/3}$);

C_h is the equivalent mass of horizontal elongated charge acting at the observation point in the ground massif, and determined by relationship (4.6).

As the seismic effect of a blast depends, largely on the extent of ground saturation with water, the coefficient k_1 can also be determined based on the moisture content in clayey soil by the formula

$$k_1 = 40e^{0.06w},$$

where w is ground moisture content by weight, ranging from 11 to 26%.

It should be noted that charges having a ratio of charge length to diameter not less than 125 are referred to as horizontal elongated charges.

Formula (4.7) is applicable to the velocity of ground movement only soils in which charges have been placed. If the structures to be protected are situated in soils having properties that differ from those in which the blast occurs, then this difference is accounted for by introducing in formula (4.7) the following factor

$$k_{h \cdot p} = \sqrt{\rho v_P / (\rho v_P)_H},$$

where the index H is related to soil properties at the point of observation.

To what extent the proper determination of intensity level of seismic vibrations in a blast of cylindrical charge as well as horizontal elongated charge is important can be seen from Fig. 39. This figure depicts the relationships $u_R^z = f(r)$ in blasts of (1) horizontal and (2) concentrated charges in loamy soils (charge weight $C = 6$ kg), and (3) similar relationship in an industrial blast of horizontal charge with a total weight 25 T. The curve corresponding to the computed data based on the formula for concentrated charge for the same industrial blast of 25 T charge up to epicentral distances equal to 6 l_c, is shown by a dashed line in Fig. 39.

As can be seen in Fig. 39, in the first case the intensity of vibrations in a blast of concentrated charge exceeds by 1.45 times the analogous indicators in a blast of horizontal charge at similar distances, while in the second case the error

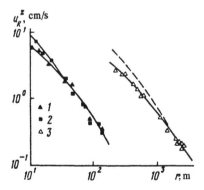

Fig. 39: Dependence of velocities of ground movement on distances in loamy soils:

1—horizontal charge $C = 6$ kg; $l = 6$ m; 2—concentrated charge $C = 6$ kg; 3—horizontal charge $C = 2.5 \times 10^4$ kg; $l = 300$ m.

in calculations using the formula of concentrated charge could be still more by 1.6 times.

The intensity of seismic vibrations in the axial direction of horizontal charge was 2–3 times less than the intensity at the same distances in a direction perpendicular to the axis of charge.

Most of the experimental data was obtained during the processing of seismograms pertaining to industrial blasts conducted in the construction of reclamation systems. Therefore, the aforementioned formulae can be used with full confidence in similar blasting conditions.

The nature of variation with distance in the period of vibrations in a surface wave induced by blasts of a horizontal extended charge in a direction perpendicular to the charge axis is shown in Fig. 40. It can be seen that the period of vibrations increased with an increase in distance as well as charge weight and also depended on soil properties. In a general form,

$$T_R = k_h \, C^{0.145} \, r^{0.05}, \qquad \qquad \dots (4.8)$$

where k_h is the coefficient that takes into account soil properties (for loams depending on their moisture content
$k_h = 0.085–0.105$; for loess $k_h = 0.055$, for fine dry cemented sand $k_h = 0.045$)

Comparison of duration of vibrations in the axial direction of charge and in a direction perpendicular to it shows that the duration in the former is more than that in the latter (Fig. 41).

4.4 Ensuring Seismic Safety of Structures during Blasts of Elongated Cylindrical Charges

A building can be damaged due to blast effects when stresses generated in the structure exceed the permissible values. However, the magnitude of stresses

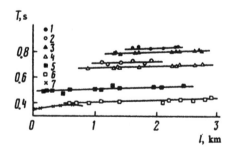

Fig. 40: Dependence of period of ground vibrations T in a surface wave on the distance l in blasts of horizontal elongated charges in a direction perpendicular to the charge:

1, 2—No. 4; 3, 4—No. 6; 5—No. 2; 6—No. 3; 7—No. 7.

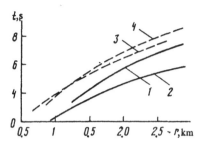

Fig. 41: Dependence of duration of ground vibrations t on distances r in blasts of horizontal extended charges in loams:

1 — $C = 11 \times 10^4$ kg, $l = 750$ m and 2 — $C = 4.7 \times 10^4$ kg, $l = 550$ m in a direction perpendicular to axis of the charge; 3 and 4 — in a direction along the axis of these charges.

developed in the structure depends not only on the nature of the seismic waves, but also on the characteristics of the structure. The period of natural vibrations T_0 in the structure and its damping properties expressed through the coefficient of attenuation of natural vibrations in buildings ε_0 are major factors.

The intensity of swaying of buildings depends on the ratio T/T_0, apart from other factors. If T is very small compared to T_0, the building will show almost no sway. At T almost equal to T_0, the vibration amplitude of buildings attains maximal value and may even exceed the amplitude of ground movement several times.

The intensity of vibrations of buildings at various ratios of T and T_0 is characterised by a dynamic coefficient β, which shows the number of times, the amplitude of vibrations in buildings exceeds the amplitude of ground movements. It is determined by the formula

$$\beta = \frac{1}{\sqrt{(1 - T_0^2/T^2)^2 + \varepsilon_0^2 T_0^2/T^2}}.$$

Through an analysis of results of blasts in dense media, associated with mining activities, a considerably lower effect of seismoblast waves on the buildings was established when compared to the effect of earthquakes. This is explained thus: the periods of ground vibrations induced by the aforesaid blasts are considerably lower than the periods of natural vibrations of most buildings and the duration of effect on the buildings is very short.

In mining enterprises, while predicting the seismic effects due to a blast, the dynamic coefficient of a building is not always taken into account.

In industrial blasts of horizontal elongated ground charges, an altogether different picture is noticed, as evident from Figs. 30, 40 and 41: forced cycles of ground vibrations reach 0.4–0.9 s, which corresponds to natural vibrations of many structures, and the duration of vibrations is about 2–9 s. Therefore, in this case consideration of time-related features of forced and natural vibrations of structures is of vital importance.

For rigid ($T_0 < 0.4$ s) and semi-rigid ($0.4 \leq T_0 \leq 0.6$) structures, the dynamic coefficient of structures during blasts is given by

$$\beta = 0.15/(\lambda^{0.4}T_0),$$

and for flexible ($T_0 > 0.6$ s) as per formula

$$\beta = 0.25/(\lambda^{0.4}T_0),$$

where λ is the decrement in damping of rock vibrations, determined by analysing seismograms.

The force acting in the section of a structure when seismic vibrations act on it, can be quantified by using the dynamic coefficient

$$S_k = k_c\beta\eta_K \cdot Q,$$

where k_c is the seismicity coefficient equal to the ratio of acceleration of ground vibrations to acceleration due to gravity;

η_K is the coefficient of shape deformation of the structure;

Q is the mass of the system.

Thus, the extent of damaging effect of blasts on buildings and structures is mainly determined by displacement velocity at the foundation of the structure; ratio of forced and natural periods of vibrations of buildings and durations (recurrence) of effect.

Structures are damaged only in such cases when the displacement velocity u exceeds a particular permissible quantity u_{per} characteristic to structures of a given type.

Apart from the permissible velocity of ground movement, the ultimate velocity is used as the criterion of intensity of vibrations. This ultimate velocity corresponds to threshold conditions of safety of structures and is used in single blasts conducted very near to a structure. The numerical difference between permissible and ultimate velocities is determined by the ratio between calculated normative and ultimate values of strength and stability, which lie in the range 0.5–0.8.

Prof. V.N. Mosinets [24] developed a methodology for seismic-safe conduction of very large blasts in mines. Using this methodology, the seismic-safe distances and weights of charges in blasts of extended cylindrical type were determined. The permissible velocity of ground vibrations was the basis for calculating the stability of structures to be protected. This depends on the purpose and state of buildings and structures of a particular class. In accordance with construction standards SNiP-II-A.3-62 and SNiP-II-A.12-69, industrial buildings and structures can be grouped into four classes:

Class I — Very important buildings and structures (All-Union and for the Republics), historical and architectural memorials, in whose vicinity conduction of blasting operations is permitted only exceptionally (Gosstroiy of the USSR or of the Republics have the authority to categorise buildings in this class).

Class II — Very important huge industrial structures (pipelines, plant buildings, mining headframes, water tanks of 20–30 years life, residential/office buildings in which a number of people live/work, apartments, cinemas, theatres, houses of culture etc.).

Class III — Industrial and service buildings of smaller dimensions (having no more than three storeys): workshops, compressor houses, civilian buildings in which fewer people live and work, apartments, shops service centres etc.

Class IV — Civilian and industrial buildings housing expensive and valuable machinery and instruments, damage to which could harm the life and health of people, godowns, service centres for transport, cold storages, compressor installations etc.

If in a group of buildings there are several structures belonging to various classes, then the permissible velocity is taken according to the most significant type of building (or structure) or according to the most damaged among them so that the minimum velocity can be selected.

The permissible velocities of ground vibrations at the foundations of buildings and structures, based on their class affiliation, are given in Table 10.

Data from the present experimental as well as industrial investigations established the seismic-safe distances and weights of charges through empirical formulae based on the criterion of mass velocities in blasts of single charges.

1. In the near-field zone as per the effect of direct longitudinal wave in blasts of vertical cylindrical charges in Kiev loam (soil characteristics are given in Table 7, site No. 5):

$$r_c = (72HC_1^{6.5}/u_{per})^{3/4}; \qquad \ldots (4.9)$$

$$C_1 = [u_{per}r^{1.3}/(65H^{0.91})]^{0.85}. \qquad \ldots (4.10)$$

2. In the far-field zone (up to distances $r = 6 \, l_c$ as per the effect of surface wave in blasts of vertical cylindrical charges in similar soil conditions (as detailed in point 1):

$$r_c = 25H_{oc}^{1/2}C^{1/3}u_{per}^{-0.8}; \qquad \ldots (4.11)$$

$$C_{per,c} = [u_{per}r^{1/3}/(70H_{oc}^{2/3})]^{2.33}. \qquad \ldots (4.12)$$

3. According to the effect of surface wave in blasts of concentrated charges:

$$r_{ccon} = \left(k_1 H_{ocon}^{1/3}C^{1/2}/u_{per}\right)^{3/4}; \qquad \ldots (4.13)$$

Table 10

Characteristics of buildings and structures	Permissible velocity of ground vibration, cm/s, as per classes of the structures		
	II	III	IV
Buildings and structures intended for civilian or industrial purposes, with reinforced concrete or metallic carcase and having antiseismic reinforcement. There are no residual deformations in the carrying elements, in the constructions and in the filler material.	5	7	10
Buildings and structures with reinforced or metallic carcase without antiseismic safeguards. No residual deformations.	2	5	7
Buildings with carcase, filler-bricks or stone construction, cracks in the filling material. New or old large-blocked or brick building without antiseismic safeguards.	1.5	3	5
Carcase type buildings, having considerable damages in the filling material and cracks in the carcase. New or old large-block or brick building, having small individual cracks in the carrying walls and barricades. New or old building of carcase type having cracks in the carcase, damaged links between individual elements.	1	2	3
Large-block or brick building with carrying walls, having considerable damages in the shape of oblique cracks, cracks in the corners etc. Buildings and structures with damage in the reinforced concrete carcase, corrosion in the armature carcase, large cracks in the filling material.	0.5	1	2
Buildings with load-carrying walls, having a large number of cracks, broken links between the external and internal walls etc. Large-panelled buildings without antiseismic safeguards.	0.3	0.5	1

$$C_{per,con} = \left[u_{per} r^{4/3} / \left(k_1 H_{ocon}^{1/3}\right)\right]^{7/3}. \qquad \dots (4.14)$$

The values of u_{per} in formulae (4.9) to (4.14) are taken from Table 10, and the values of k_1 from Table 7.

H_{ocon} is the scaled depth of placing the concentrated charge, $\text{m/kg}^{1/3}$.

If the blasting conditions remain constant (i.e., blasting around the same structure intended to be protected, at the same epicentral distances, but not exceeding 6 l_c) and if there exists the technological possibility or compulsion to conduct the same blast with a horizontal extended charge, then the total permissible weight of the horizontal extended charge would considerably exceed the weight of concentrated charge, and is determined by the formula

$$C_{per,h} = C_{per} / (0.43 e^{0.15 r l_c^{-1}}). \qquad \dots (4.15)$$

For example, in blasting a concentrated charge weighing 4.4 T in loams, the seismic-safe distance for the structure, at $u_{per} = 1.5$ cm/s, would be 300 m. Without increasing r_c and without changing the conditions of blasting, a horizontal extended charge of 300 m length weighing 9 T could be blasted.

A universal nomogram is shown in Fig. 42, which allows determination of the total permissible weight of concentrated and horizontal extended charges or any other unknown blasting parameter for blasts in the soils enumerated in Table 7.

To use the nomogram, initially the permissible velocity of vibrations for the structure has to be determined, the coefficient k_1, which depends on soil properties, has to be established, the scaled depth of charge placement has to be determined and for the given distance between blast site and the structure, the total permissible weight of charge should be ascertained.

An example is shown (dashed line) in the nomogram for determining the total permissible weight of concentrated and horizontal extended charges for the following parameters:

$$u_{per} = 2 \text{ cm/s}; \ k_1 = 75; \ H_0 = 0.6 \text{ m/kg}^{1/3}; r = 600 \text{ m}.$$

Permissible weight of concentrated charge 70 T.

If it is necessary to blast a horizontal extended charge of length $l_c = 300$ m for the same conditions and parameters mentioned above, then taking into account the distance up to the structure to be protected $r = 600$ m and the ratio $r/l_c = 2$, the permissible weight of the horizontal extended charge is ascertained from the nomogram to be equal to 120 T.

From the investigations into the seismic effects induced by a blast of extended horizontal charge, it is evident that in cases when rigid directionality in the charge axis is not required by technological conditions, the seismic effects of a blast on the structure being protected can be reduced by 2–3 times or even

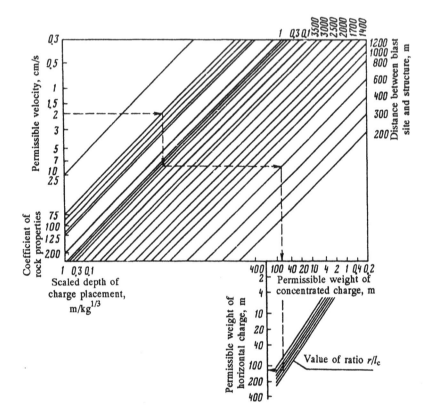

Fig. 42: Nomogram for determining the permissible weight of concentrated and horizontal extended charges.

more (depending on the construction of charge and distances up to the structure), if it is directed towards the structure.

5

Reducing the Magnitude of Blast-induced Seismic Wave Effects by Using Artificial Screens

5.1 Propagation of Elastic Wave Through a Screen

The objective of the present study is to obtain the coefficient of screening effect and to establish its dependence on parameters of the screen and disturbances incident upon it. The principle of screening is determined by the conditions of wave propagation through a medium distinctly different in property compared to the surrounding massif.

We may seek a theoretical solution to the problem of the mutual effect between a non-stationary wave and an interlayer of given thickness d_W (Fig. 43) and wave propagation velocity v_2 [18]. Let us establish the laws of wave propagation excited in the system by a source, assuming that the wave process is described by the functions $a_1(x, z, t), a_2(x, z, t)$ and $a_3(x, z, t)$. Satisfying the conditions (outside the source) in media $z > 0$, $d_W < z < 0$ and $z < -d_W$ and the following equations correspondingly

$$\Delta a_i = b^2 \frac{\partial^2 a_i}{\partial t^2};$$

$$\Delta a_2 = c^2 \frac{\partial^2 a_2}{\partial t^2}; \quad (i = 1; 3),$$

where b is a quantity inversely proportional to propagation velocities of waves in media I and III, $b = 1/v_1$;

c is the same as b, but in medium II, $c = 1/v_2$.

Threshold values at $t = 0$

$$a_i|_{t=0} = \frac{\partial a_i}{\partial t}|_{t=0} = a_2|_{t=0} = \frac{\partial a_2}{\partial t}|_{t=0} = 0 \qquad \dots (5.1)$$

are subject to the following conditions at the boundaries separating the media:

$$a_1 = a_2, \frac{\partial a_1}{\partial z} = \frac{\partial a_2}{\partial z} \quad \text{at } z = 0;$$

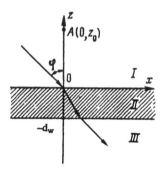

Fig. 43: Diagram for determining the effectiveness of artificial screens:
I, III — media; II — interlayer.

$$a_2 = a_3, \frac{\partial a_2}{\partial z} = \frac{\partial a_3}{\partial z} \quad \text{at } z = -d_W. \qquad \dots (5.2)$$

A plane wave may incident at an angle φ. Let us assume that the disturbance corresponding to the incident wave is represented by the integral

$$a_0(x, z, t) = \frac{1}{2\pi j} \int\limits_{(l)} S_0(p) e^{p(t-\alpha bx + \beta bz)} dp,$$

where $S_0(p)$ is the complex function of variable p (spectrum of incident disturbance); $\alpha = \sin \varphi; \beta = \cos \varphi$.

At $S_0(p) = \dfrac{1}{(p - j\omega)}$ the incident disturbance would be a plane monochromatic wave, $a_0 = e^{j\omega(t-\alpha bx+\beta bz)}$.

If, at $S_0(p) = \dfrac{1}{p}$, we integrate the linear function $R_e \cdot p = \sigma > 0$, then the incident disturbance takes the shape

$$a_0 = \begin{cases} 1 \text{ at } t > \alpha bx - \beta bz; \\ 0 \text{ at } t < \alpha bx - \beta bz. \end{cases}$$

Let us find the functions a_1, a_2 and a_3 in the form of integrals,

$$\left.\begin{aligned}
a_1 &= a_0 + \frac{1}{2\pi j} \int\limits_l S(p) e^{p(t-\alpha bx-\beta bz)} dp; \\
a_2 &= \frac{1}{2\pi j} \int\limits_l [S_1(p) e^{p\beta cz} + S_2(p) e^{-p\beta_2 cz}] e^{p(t-\alpha_2 cx)} dp; \\
a_3 &= \frac{1}{2\pi j} \int\limits_l S_3(p) e^{p(t-\alpha_3 bx+\beta_3 bz)} dp;
\end{aligned}\right\} \qquad \dots (5.3)$$

where $\alpha_2 = \alpha/\gamma; \alpha_3 = \alpha;$

$$\beta = \beta_3 = \sqrt{1 - \alpha^2};$$

$$\alpha = \sin\varphi; \quad \gamma = \frac{c}{b} < 1; \quad \beta_2 = \sqrt{1 - \frac{\alpha^2}{\gamma^2}}.$$

Using conditions (5.1) and (5.2) the coefficients are established

$$S(p) = \frac{1 - \gamma^2}{\Delta(p)} S_0(p)(1 - e^{2pc\beta_2 d_w});$$

$$S_1(p) = -\frac{2\beta S_0(p)}{\Delta p}(\gamma\beta_2 + \beta)e^{2pc\beta_2 d_w};$$

$$S_2(p) = -\frac{2\beta}{\Delta(p)}(\gamma\beta_2 - \beta)S_0(p);$$

$$S_3(p) = S_0(p)e^{(b\beta - c\beta_2)pd_w}\left[1 - \frac{(\beta - \gamma\beta_2)^2}{\Delta(p)}\left(1 - e^{-2pc\beta_2 d_w}\right)\right],$$

where $\Delta(p) = (\beta - \gamma\beta_2)^2 - (\beta + \gamma\beta_2)^2 \, e^{2pc\beta_2 d_w}$.   ... (5.4)

Subsequently our interest shifts to the disturbance $a_3(x, z, t)$ that become refracted in medium, III.

Let us consider the case $S_0(p) = \dfrac{1}{(p - j\omega)}$ and select a smaller circle with its centre at $p = j\omega$ as the contour l for integration. The last integral in (5.3) is calculated using the theorem of residues considered in the theory of a complex variable. As per this theorem, we find the residual at point $p = j\omega$,

$$a_3 = e^{j\omega d_w(b\beta - c\beta_2)}\left[1 - \frac{(\beta - \gamma\beta_2)^2}{\Delta(j\omega)}\left(1 - e^{2j\omega d_w c\beta_2}\right)\right] \times$$

$$\times \, e^{j\omega(t - \alpha bx + \beta bz)}.$$

Taking into account (5.4) and putting $\beta_2 = -j|\beta_2|$, we find

$$A = 1 - \frac{(\beta + j\gamma|\beta_2|)^2\left(1 - e^{-2\omega d_w c|\beta_2|}\right)}{(\beta - j\gamma|\beta_2|)^2 - (\beta + j\gamma|\beta_2|)^2 \, e^{2\omega d_w c|\beta_2|}}.$$

Thus, the plane wave passing through the screening interlayer depends both on the parameters of incident disturbance and on the screen parameters. Let us consider the quantity a_3/a_0 and term it as a screening coefficient. Its value for the above given screen is determined by the formula

$$\eta = \frac{a_3}{a_0} = Ae^{j\omega(b\beta - c|\beta_2|)d_w}.$$ ... (5.5)

After separating the real part in the complex expression (5.5), we obtain

$$\eta = \frac{4\gamma\beta\beta_2(\alpha_1 \cos \omega d_{\mathrm{w}}\overline{m} - \alpha_2 \sin \omega d_{\mathrm{w}}m)}{\alpha_1^2 + \alpha_2^2}, \qquad \ldots (5.6)$$

where $\alpha_1 = n_1^2 \cos 2\omega d_{\mathrm{w}} c\beta_2 - n_2^2$;

$\alpha_2 = n_1^2 \sin 2\omega d_{\mathrm{w}} c\beta_2$;

$n_1 = \beta + \gamma\beta_2$;

$n_2 = \beta - \gamma\beta_2$;

$m = c\beta_2 - b\beta.$

Let us study the dependence of the screening coefficient on the parameters $d_{\mathrm{w}}, \omega, \gamma$ and $d_{\mathrm{w}}/\lambda_{\mathrm{w}}$ (ratio of screen width or cavity diameter to the wavelength of the incident disturbance).

At $\varphi = 0, \gamma = 3, v_1 = 1/b = 600$ m/s, we have

$$\frac{a_3}{a_0} = \cos 0.1 d_{\mathrm{w}} - \frac{4 \cos 0.1 d_{\mathrm{w}} - 4 \cos 0.3 d_{\mathrm{w}}}{17 - 8 \cos 0.2 d_{\mathrm{w}}}.$$

At $d_{\mathrm{w}} = 1$ m and $v_1 > v_2$, the dependence of coefficient η on velocity of propagation γ of elastic wave through the screening interlayer and frequency of incident disturbance $\log \omega$ are shown in Fig. 44.

At $\varphi = 0$, $\gamma = 3$, by means of transformation, we obtain the relationship

$$\frac{a_3}{a_0} = \cos 18.84 \frac{d_{\mathrm{w}}}{\lambda_{\mathrm{w}}} - \frac{4 \cos 18.84 d_{\mathrm{w}}/\lambda_{\mathrm{w}} - 4 \cos 56.52 d_{\mathrm{w}}/\lambda_{\mathrm{w}}}{17 - 8 \cos 37.68 d_{\mathrm{w}}/\lambda_{\mathrm{w}}}.$$

The relationships between the coefficient of screening and the ratio $\lambda_{\mathrm{w}}/d_{\mathrm{w}}$ are obtained on the basis of aforementioned theoretical solution to the problem of propagation of an elastic wave through an interlayer and the experimental data. It is evident that theoretical data agree satisfactorily with the experimental.

5.2 Results of Experimental Investigations

The maximum power of a blast is limited by the radii of seismic-safe distances, especially in areas hosting various buildings and structures. These restrictions constrain the enlargement of industrial operations, affecting labour productivity and call for safety measures in the conduction of blasting operations. All these aspects reduce the total effectiveness of utilising the blast energy in different fields of the national economy [20].

A set of methods to reduce the seismic effects of a blast is available. It is well known that the intensity of blast waves generated under the influence of a completely formed pulse during the course of their propagation, is determined only by the properties of the medium and distance from the source of disturbance. In such a case the intensity of blast waves can be controlled only through the means of artificial interventions on the path of wave propagation.

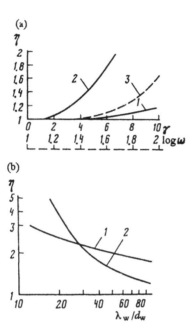

Fig. 44: Dependence of the screening coefficient on:

a— velocity of propagation of elastic wave through the screening interlayer (1 —angular frequency $20^{-1}s^{-1}$; 2 —same, $63^{-1}s^{-1}$; 3 —frequency of incident disturbance);

b— ratio of wavelength to width of screen (1 —theoretical calculation; 2 —experimental data).

One such method involves the creation of various artificial screens on the propagation path of a blast wave. These screens could be set up either mechanically or through blasting methods. They could either be continuous or discontinuous. The latter type (through blasting and of a discontinuous nature) involves drilling holes with a specified spacing and subsequent charging and blasting a row of such holes. Depending on the site-specific conditions, blasting could be done hole-wise or completely grid-wise (if needed, with delays between holes). By blasting such holes, we obtain cavities of the required cross-section and around them zones of plastic deformations and zones of compacted soil in which a disturbed structure or discontinuity exists. Based on the parameters of cavities, zones of plasticity and compaction (with charge diameter calculated accordingly), the spacing between holes in the row is selected.

Reduction in the intensity of seismic waves induced by blasting by means of the screen is achieved in the following manner. As a seismic wave approaches the screen, a part of its energy is reflected in the massif (owing to the interface 'cavity-medium'). The quantity of reflected energy depends on the ratio of acoustic rigidities of the soil and air.

The wave energy dissipates significantly even in the gaps between cavities because of propagation through the fissured and loosened layer. Therefore a further reduction in the intensity can be expected.

There are two methods for erecting such artificial screens: (1) nearer to the source of disturbances and (2) directly in front of the unit to be protected. In the first method, the screen is temporary.

In multiple blasts, and sometimes even in single blasts, it is advisable to form a screen directly in front of the protected building. In such a case the screen is set up (formed) on a permanent basis and may be made using an absorbent material (Kermasite). The screen may consist of a single row or multirows and the cavities of seismic screens may be vertical, and if need be even inclined with an enlarged upper part or lower part (Fig. 45).

The effectiveness of artificial screens in the shape of a row of cavities was studied in the laboratory and through prototype models as well as in field conditions.

Reduction in the intensity of elastic wave beyond the screen—depending on varying diameter of cavities, spacing of cavities in a row, distance between screen and source of waves or protected structure—was studied by modelling in the laboratory. This enabled a study to be made of the effectiveness of artificial screens in real conditions.

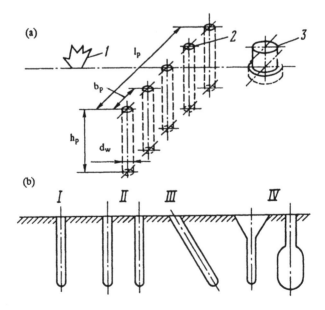

Fig. 45: (a) General technological layouts for installing and (b) types of artificial screen:

1—blast site; 2—screen; 3—protected unit; I—single row screen; II—multirow screen; III—screen with inclined pits; IV—screens with enlargement of collars or lower parts of the pit.

Table 11

A_{01}, mm	A_{02}, mm	A_{03}, mm				
6	14	30	b, mm	η_1	η_2	η_3
A_1, mm	A_2, mm	A_3, mm				
1.5	4	4.5	10	4	3.5	6.7
2	6	10	12	3	2.34	3
3	6	18	14	2	2.34	1.67
3	8	18	16	2	1.75	1.67
4	9	22.5	18	1.5	1.56	1.34
4	9.5	21	20	1.5	1.48	1.43
4	10	20	22	1.5	1.4	1.5
4	11	23	24	1.5	1.27	1.3
5	11	26	26	1.2	1.27	1.15
4	10	24	28	1.5	1.4	1.25
4	9	24	30	1.5	1.56	1.25
4.5	11	25	32	1.33	1.27	1.2
4	10	24	34	1.5	1.4	1.25
5	11	27	36	1.2	1.27	1.11
5	11	27	38	1.2	1.27	1.11
4.5	11.5	21.5	40	1.33	1.32	1.39
5	11.5	27	26	1.2	1.22	1.39

Note: A_{0i}, A_{02}, A_{03} — amplitudes of waves without a screen (test waves); A_{01}, A_{02}, A_{03} — amplitudes of waves in the presence of a screen. Screen length along wavefront $l_p = 280$ mm.

Some results of these investigations on the intensity of a wave field at 6 mm diameter of cavities are given in Table 11. The coefficient of reduction in wave intensity η varied between 1.1 and 6.5 depending on parameters of the screen. In experiments conducted with the diameter of cavities approximately equal to the wavelength of the source, the coefficient η was found to increase. This confirms the view on the increase in the effectiveness of the screen for comparable dimensions of cavity diameter and the length of approaching wavefront.

As can be seen from Fig. 46, the spacing between cavities considerably influences the variation in intensity of waves passing through the screen. In the experiments conducted the optimal spacing was equal to twice the diameter of cavity, while at a spacing of four times the diameter the influence of the screen on the intensity passing wavefront was practically insignificant.

Thus the model experiments confirmed the possibility of decreasing the intensity of vibrations in a wave by installing artificial screens in the path of propagation.

Prototype modelling of artificial screens was done at the Experimental Blasting Base of Academy of Sciences, Ukrainian SSR in order to establish the optimal parameters. Modelling was done by recording the blast wave parameters with the help of standard seismometers VEGIK along with galvanometers

104

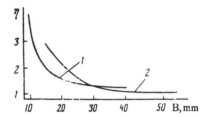

Fig. 46: Variation in intensity of vibrations in wave η based on spacing of cavities in screen b:
1—diameter of cavities, 6 mm; 2—diameter of cavities, 8 mm.

MOO 1.2. These enabled quantifying the rate of soil particle displacement in the oscillogram.

The measurement profiles of displacement velocities were laid according to a single ray pattern as well as multiray pattern so as to fix the 'shadow' of the screen.

Just as in the case of laboratory modelling, the reduction coefficient in the intensity of vibrations η was taken to be the ratio of soil particle velocities in the wave before reaching the artificial screen and behind it.

Parameters of the screen were: diameter of cavities d_w or, in other words, width of partition; spacing of cavities in a row, b; its length l and depth h.

During the experiments, 88 trial blasts were conducted. In each case, the screen was formed by means of blasting. Each screen served for conducting 3–4 experiments. Charges of 0.2–5 kg were placed at 1–4.5 m depth. Partition parameters were: diameter of cavities 0.11–0.35 m, spacing of cavities in a row 0.8–2 m, length 9–22.5 m and depth 2–4.5 m.

As far as possible, the blasts were conducted with the same layout of sensors, which avoided errors in measurement associated with the workmanship of installing the sensors.

Two influencing factors on the intensity of blast waves were distinctly observed in front of the screens and behind them: reduction in their intensity due to the presence of pits and, on the other hand, increased intensity due to the passing of waves through a zone of residual soil deformation, created during the formation of the screen by blasting. As the wave approached this zone the particle velocity increased and, after passing the row of pits in the screen, it dropped, reaching the maximum coefficient of reduction $\eta = 2.85$. Further, the velocities approached to those values of wave parameters observed in the test blast. The field of equal magnitudes of velocities of the presently considered blast and test blasts was situated at a distance $r_0 = 50$, i.e., the size of the 'shadow' behind the screen (effective zone of artificial screen) was equal to 55 mm or 2.5 l.

It was observed from experiments that the dimensions of the 'shadow' depend on the dimensions of the screen along the wavefront as well as the depth of the massif, and also on the power of blast since the period of vibration or wavelength is related to the latter.

Blast No. 40 can be taken as an example. A charge of alumotol weighing 8 kg in a hole of 2 m was blasted in this experiment. A cavity was made at a distance of 5 m from the charge. The parameters of the cavity were: spacing in a row 1.5 m, depth 2 m, length 10 m. The measurement profile was laid out according to the three-ray pattern. The last external ray was oriented towards the charge axis and the extreme pit in the cavity. The isolines of particle velocities within the zone and outside the zone of action of artificial screen, were drawn after processing the experimental data. If we superimpose on such a graph contours of velocities of a similar blast conducted without an artificial screen, the points of intersection of these contours will form the screening 'shadow' of the cavity.

As can be seen from Fig. 47, within the zone of influence of a cavity, in a particular section in which the soil particle velocity along z-component decreases by three times its dimensions are less than the dimensions of that section situated outside the effective zone of cavity by 2.2 times. This effect is felt up to relative distances of $r_0 = 13$, i.e., in this case the effective zone of screen extends up to distances equal to 3 l. This is explained by the fact that the effectiveness of a screen increases to a certain extent when installed directly in front of the charge.

Figure 48 shows the variation in soil particle velocities depending on distances up to the screen as well as on the influence of the latter (in quantities, reduced to charge mass) in blasts of several charges of different weights and located at different distances from the screen. The length of the screen along the wavefront was 9 m (points 1, 2) and 22.5 m (points 3–6) and width 0.35 m. Analysis of the data given in Fig. 48 and Table 12 once again confirmed that the effective zone of the screen extended up to 2.5 l.

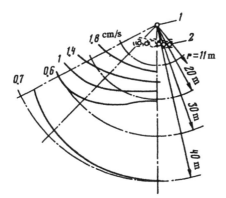

Fig. 47: Effective zone of artificial screen:

1—Explosive charge; 2—screen.

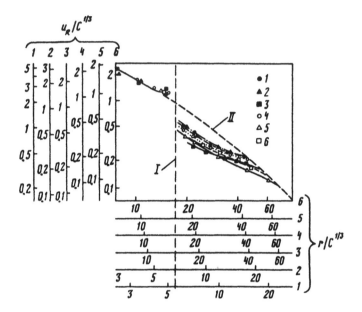

Fig. 48: Dependence of soil particle velocities on distance up to the screen and screen effectiveness for different charge weights:

1 — 20 kg; 2 — 50 kg; 3 — 5 kg; 4 — 12 kg; 5 — 0.8 kg; 6 — 1.5 kg. I — Screen location; II — relationship $u = f(r)$ in test blasts.

While studying the effect of artificial screens on the parameters of blast-induced seismic waves, a relationship was noticed between the period of vibrations in the incident wave, diameter of pits at the screen and effectiveness of screen. Reduction in the period of vibrations, i.e., wavelength, at unchanged parameters of the screen, leads to an increase in the value of coefficient η:

$$\eta = 1.34(d_W/T)^{0.22}. \qquad \ldots (5.7)$$

From formulae (4.1) to (4.3) and from data from other sources it can be seen that the period of vibrations depends on charge weight, epicentral distances and soil conditions. The optimal diameter of pits in the screen that ensures stability of the protected structure can be established using these relationships.

The artificial screen was placed in front of the charge (Table 13) and in front of the protected structure (Table 14) so as to study the effect of spacing of drill holes on the velocity of soil particles in the incident wave. The charge weight remained the same (14 kg) in both cases while the distance between drill holes in the screen varied from 1 to 2 m. It was noticed that the best effect in reducing the particle velocity in the case of the screen placed in front of the charge was achieved at 1 m spacing of drill holes, i.e., $3d_W$, while with the screen located in front of the protected structure, the spacing of drill holes had

Table 12

No. of blasts	Period of vibrations in wave before meeting screen T, s	Velocity of displacements in wave in front of screen $u_R^z/C^{1/3}$, cm/s	Screen length along wavefront l, m	Width of screen (diameter of pits) d_w, m	Velocity of displacements behind the screen $u_R^z/C^{1/3}$, cm/s	Extent of effective zone of screen, L, m	Screening coefficient, η
46	0.1	0.48	22.5	0.35	0.28	57	1.7
49	0.09	0.55	22.5	0.35	0.31	44	1.78
48	0.1	0.68	22.5	0.35	0.38	55	1.78
47	0.11	0.63	22.5	0.35	0.37	57	1.68
52	0.07	0.51	22.5	0.35	0.26	46	1.97
51	0.08	0.87	22.5	0.35	0.47	52	1.85
56	0.07	1.76	22.5	0.35	0.62	55	2.85
20	0.15	1.83	22.5	0.35	1.2	25	1.52
34a	0.06	0.25	20	0.11	0.16	42	1.52
49a	0.07	0.34	20	0.11	0.23	44	1.47
50a	0.08	0.44	20	0.11	0.3	55	1.44
55a	0.1	0.55	20	0.11	0.39	44	1.4

Table 13

Distance between charge and screen	Spacing between drill holes in the screen, m	Particle velocities at scaled distances along z- and x-components, cm/s					
		$r_0 = 10$ m		$r_0 = 15$ m		$r_0 = 30$ m	
		z	x	z	x	z	x
3.3	1	2.5	2.8	—	—	0.6	0.8
3.3	1.5	4.5	2.8	—	—	1.2	0.79
3.3	2	5	2.8	2.5	1.38	0.5	0.4
5.4	1	1.3	2.25	—	—	0.63	0.76
5.4	1.5	3.9	2.4	—	—	1.26	0.69
5.4	2	5.25	3.4	2.5	1.7	0.5	0.63

Note: Width of screen = 0.35 m.

much less impact on the variation in the parameters of blast-induced seismic waves. Nevertheless the favourable spacing was 2 m, i.e., 5.5 d_w.

In prototype modelling, the spacing of drill holes increased compared to laboratory modelling for the same values of η. This is explained by the existence of discontinuities between the pits in the screen owing to crack formation and other deformations.

Curves characterising the change in velocity of the surface wave over distance up to the screen for three blasts using charges of $C = 14$ kg are shown in Fig. 49. In blast No. 1, the artificial screen was positioned at 13 m from the

Table 14

Spacing between drill holes in the screen, m	Particles velocities at scaled distances along z- and x-components, cm/s			
	$r_0 = 24$ m		$r_0 = 28$ m	
	z	x	z	x
1	0.66	0.66	0.43	0.3
1.5	0.57	0.58	0.52	0.42
2	0.56	0.54	0.45	0.3

Note: Distance from charge to screen 20 m, width of screen 0.35 m.

charge. It was well beyond the boundary of the zone of residual soil deformations, whose dimensions for the blasts under consideration was 9.5 m. In blast No. 2, the screen was placed at the boundary of the residual deformation zone (at a distance of 8 m from the charge). As can be seen, apart from lessening of wave intensity compared to the test blast (by 1.6 times), a significant reduction occurred in soil particle velocities when the screen was placed at the boundary of the zone of residual soil deformations (by 1.4 times additionally). This determines the optimal point at which the screen needs to be placed in front of the charge.

It should be noted that any increase in the distance between the blast site and the site of screen, with the parameters of the latter remaining the same, results in a reduction in the value of coefficient η beyond the screen. This occurs due to the increased period of vibrations with distance and also due to

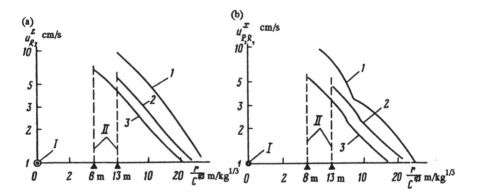

Fig. 49: Change in the velocity of surface wave over distance up to the screen which is placed in the path of propagation (a) along z-component and (b) along x-component.

1—test blast (without screen); 2—blast No. 1; 3—blast No. 2; I—charge location; II—screen location.

the diffracted waves. For example, the maximum values of coefficients (η) were obtained when the screen was placed at distances 5–35 m from the blast site with charges weighing 3, 8 and 12 kg (values of η 2.32, 2.08, 1.88 respectively) in the boundary region of the zone of residual soil deformations, while minimum values were obtained when the screen was situated at a distance of 35 m (values of η 1.75, 1.55 and 1.45 respectively).

While determining the depth of pits in the screen, the peculiarities of propagation of body and surface waves through the soil massif with increase in depth from surface and propagation of different waves through the screen should be taken into consideration.

The present studies and other investigations into the propagation of blast waves through a soil massif revealed an intensive damping of surface waves with increase in depth and a reduction in amplitudes of vibrations in the body wave with increase in distance from the free face.

Quantitative features of the longitudinal and surface waves within the soil massif gave rise to the relationships shown in Fig. 50 and the following formulae were obtained by the authors and other researchers.

For loess-type loam

$$\frac{a_z(h_p)}{a_z(0)} = 1.34e^{-2h_p/\lambda_w} - 0.34e^{-5.95h_p/\lambda_w};$$

$$\frac{a_x(h_p)}{a_x(0)} = 0.42e^{-2h_p/\lambda_w} - 2.02e^{-5.95h_p/\lambda_w},$$

where λ_w — wavelength.
In granite

$$\frac{a_z(h_p)}{a_z(0)} = 0.94e^{-2.31h_p/\lambda_w} + 0.06e^{-5.81h_p/\lambda_w};$$

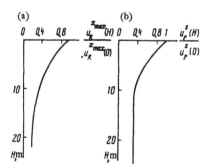

Fig. 50: Curves showing the variation of ratios u_{max}^R at depth H and at free surface 0, for $r = 275$ m: a—$u_R^x(H)/u_R^x(0)$ for blasts in loess type loam ($C = 160$ kg); b—$u_p^z(H)/u_p^z(0)$ for blasts in hard rocks ($C = 1$ to 100 T).

$$\frac{a_x(h_P)}{a_x(0)} = 0.91e^{-2.31 h_P/\lambda_w} - 1.77e^{-5.81 h_P/\lambda_w}.$$

It should further be noted that with surface waves the x-component, which is more harmful for buildings and structures, attenuates rapidly, i.e. there is the opportunity to reduce the intensity of vibrations by forming shallow screens in the path of wave propagation.

To reduce the intensity of body waves, the authors suggest that more than the depth of the screen, its slope is of paramount significance, i.e., formation of an inclined screen immediately near the protected building such that the building falls inside the 'geometrical shadow' zone of the partition.

Thus experiments confirmed the high efficiency of an artificial screen in absorbing the intensity of blast waves. In the experimental investigations and during field trials, the coefficient of reduction obtained in velocities was about 1.5–2 at the ratio $\lambda_w/d_w \approx 50$.

Considerations involved in installing an artificial screen in front of a protected object include:

(a) Selection of the main parameters of the seismic wave (velocity and frequency) near the protected object and in the area proposed for setting up the artificial screen, are selected.

(b) Determination of the permissible velocity of vibrations u_{per} for the protected object.

(c) Choosing the length of screen along the wavefront such that the protected object falls inside the 'shadow' zone of the screen $(2.5\ l)$. In this case, the depth of screen should not exceed $\dfrac{u_{per}(0)}{u(h_P)} \geq 1$.

(d) Establishing the screen width (diameter of pit d_w) from condition (5.7).

A screen can comprise several rows and field trials with artificial screens have confirmed their effectiveness.

Seismic Effects of a Blast During Formation of an Artificial Screen in Soils: Let us consider the blasting of a series of concentrated cylindrical charges in order to study the seismic effects induced by blasting a row of charges to create an artificial screen.

In the blasting of dispersed charges, blast waves from each charge in the group become superimposed at any point in a massif. Charges interact with each other, resulting in additional soil breakage and loss of part of the seismic energy.

According to studies conducted by B.G. Rulev and D.A. Kharin, an effective charge quantity C_{ef} (for dispersed charges) is established which characterises an individual charge C_1 in relation to the output of seismic energy,

$$C_{ef} = \frac{C_1}{m^{\frac{1-n}{n}}} \left[1 + \left(m^{\frac{1-n}{n}} - 1 \right) \frac{l_1}{2r_e} \right], \qquad \ldots (5.8)$$

where m is the number of charges in the group;

n is the index showing the level of damping of displacement velocity depending on the charge weight; $u = f(C^n)$;

l_1 is the distance between separate charges in the row.

Movement of soil particles at any point is determined by the superposition of waves induced by blasting of each charge. In computations, the weight of charge in each hole is taken to be equal to C_{ef}. The maximum resultant amplitude, for example displacement velocity, is observed within the boundaries of the first half-period of wave coming from the nearest blast and is calculated by the formula

$$u = KC^{n/3} \sum_{i=1}^{m} \frac{1}{r_i^n} \qquad \dots (5.9)$$

which follows from the formula for a single concentrated charge

$$u = K(r/C^{1/3})^{-n}.$$

Using the results of the aforementioned investigations on the effect of blasting single charges, it is possible to determine the intensity of vibrations when a row of concentrated charges are blasted, using equation (5.9).

The relationship between soil particle velocity and distance as well as the number of simultaneously blasted charges at the blasting site of Marganetsk Mining and Benefication Plant (Fig. 51), justify the suggested summation. Similar results were obtained while comparing the relationships obtained through experimental and computational studies on the blasting of a row of 8 cylindrical charges (Fig. 52, broken lines 1 and 3).

As can be seen from Figs. 51 and 52, data on the main criterion of 'seismic hazard-soil particle velocities' when a row of charges are simultaneously blasted (for creating a screen) determined by experimental and computational techniques according to formula (5.9) [based on the indices of blasting a single charge], differ by some 10–20%. This is acceptable for assessing the seismic effect of a blast for creating an artificial screen in front of a protected object.

Velocities computed in the case of blasting a row of 40 charges (Fig. 51, straight line 4) exceed only by 3.5–4 times the velocities in a blast with a single charge (Fig. 51, curve 1) for the same distances. This indicates the possibility of using a more effective method of constructing screens: simultaneous blasting of the entire sequence of charges in the immediate neighbourhood of protected objects. The condition for seismic safety is generally chosen from $u \leq u_{per}$. Moreover, the velocities while blasting a row of dispersed charges, can be approximately calculated according to the formula for a cylindrical horizontal charge (4.5) and seismic-safe charge weights according to formula (4.12) when a cylindrical horizontal charge is placed at the level of centre of gravity of the vertical charges in the screen and the equality condition of total mass and lengths of screen and charges are satisfied.

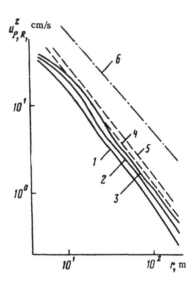

Fig. 51: Relationship between particle velocity, distance and number of simultaneously blasted charges:

1—single charge (24.5 kg); 2—two charges (21.7 and 24.4 kg); 3—three charges (26, 30 and 29 kg); 4, 5—relationships respectively for double and triple charges, calculated according to formulae (5.8) and (5.9); 6—computed relationship for screen created by blasting 40 charges, each weighing 24 kg.

A blast involving a series of vertical charges can be approximately considered as a blast of a flat charge whose vibrations attenuate to a lesser extent than in the blast of a cylindrical charge. This is confirmed by analysing the blast effects in the far-field zone.

5.3 Management of Seismic Effects Induced by Blasting

Short Delay Blasting for Compacting Collapsible Soils: Collapsible soils, mainly loess, are widely found throughout the territory of the USSR, especially in the Ukraine, northern Caucasus, Central Asia, Povolzh'e, northern Siberia, Kazakhstan and other regions. Therefore, the tasks for the design, construction and maintenance of buildings and structures built on collapsible soils acquire special significance. The technically advanced methods of building constructions create conditions for switching over to modern methods of laying foundations in collapsible soils by changing the basic physicomechanical properties of such soils in their natural habitat (place of deposition).

One perspective and advanced method is to utilise the energy of blasts for compacting soils. This method is advantageous for compacting loess-type soils since it requires relatively less water for wetting the soil mass until it attains an optimal moisture content and a short waiting time between completion of soil

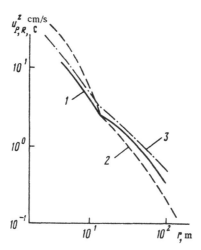

Fig. 52: Particle velocity versus distance:

1 — in a blast of a row consisting of 8 cylindrical vertical charges weighing in aggregate 8.1 kg; 2 — in a blast of horizontal cylindrical charge weighing 8 kg; 3 — relationship calculated according to formula (5.9).

compaction operations for laying foundations and the beginning of construction activities at the site. During the compaction of soils, a blast serves as a source of relatively powerful seismic vibration, which could be harmful and sometimes even damage buildings and structures situated near the construction site. In such situations, forecasting the seismic resistance of structures acquires paramount importance.

Existing engineering methods for forecasting seismic safety of a blast make it possible to establish only approximately the level of vibrations induced by blasts of scattered charges in the soil massif. Often they do not satisfy the growing demand for accuracy of prediction.

The problem of seismic safety should be solved considering the seismic resistance of buildings and structures, which in turn depends on the strength of building materials, interelement joints as well as on the dynamic characteristics of the constructions. The latter respond selectively to the vibrations, as they happen to be resonant systems.

Considering the resonance properties of buildings and structures, their seismic safety can be ensured by reducing the energy of vibrations at which resonance effects could occur. Since the vibrations are represented by a spectrum of frequencies, the seismic hazard can be reduced using the principle of energy redistribution between frequencies with such a calculation that the energy-saturated level of the resonance range in the spectrum is reduced.

The Sadovskii method is used in the USSR, primarily for determining the potential danger due to seismic effects.

The method involves the following:

—— Determining the velocity of vibrations at the foundation of the object to be protected.

—— Establishing an empirical relationship between velocity of vibrations in the soil and charge weight.

—— Assigning a selected permissible velocity of vibrations, whose seismic-safe distance is determined according to the empirical relationship (1.5).

In spite of the simplicity of assessing the bahaviour of buildings and structures with the help of displacement velocities, usage of the latter does not help in predicting the behaviour of structures over the wide frequency range of seismic vibrations. It has become evident that data on the frequency characteristics and spectrum of seismic vibrations are required in order to assess the behaviour of structures.

In the actual practice of building structures on collapsible soils, situations often arise wherein the calculated values of displacement velocities exceed the permitted values. In such cases it is advisable to adopt special methods for controlling the seismic effects [12].

The technological scheme for compacting loess-type soils with the help of blast energy envisages series blasting of a group of charges (with specific delay interval $\Delta\tau$). The delay interval is chosen in such a way that the successive charge is blasted at that moment when the blast wavefront generated by the preceding charge passes through it. The delay interval is $\Delta\tau = a_c/v$, where a_c is the distance between charges, m; and v is the velocity of propagation of elastic wave on soil, m/s.

In order to determine the influence of SDB (short delay blasting) on seismic effects, recourse is made to the principle of superposition of displacement velocities generated by blasting charges separately.

According to Sharp, displacement in the elastic zone for i^{th} charge is described by the expression

$$u_i = u_{mi}e^{-\beta_u t}\sin\omega_i t,$$

where u_{mi} is the maximum amplitude of displacement velocity for a charge.

According to the experimental data taken from several investigations

$$u_{mi} = k(C_i/r_i)^n,$$

where C_i is the mass of ith charge $(i = 1, 2, \ldots, m)$;

n is the index of order (determined experimentally).

Let us consider a case in which $C_1 = C_2 = \ldots = C_m = C$ (then $\omega_i = \omega$). We shall note that in the elastic zone distances from observation points to the individual charges are almost equal, i.e., $r_1 = r_2 = \ldots = r_m = r$.

Let us determine the resultant velocity of displacements where m number of similar charges are blasted simultaneously:

$$u_0 = k(C/r)^n \ e^{-\beta_u t} \sin \omega t + k(C/r)^n \ e^{-\beta_u t} \sin \omega t + \cdots +$$

$$+ k(C/r)^n \ e^{-\beta_u t} \sin \omega t = nk(C/r)^n \ e^{-\beta_u t} \sin \omega t. \quad \cdots \ (5.10)$$

In the case of SDB, with delay interval $\Delta\tau$, we have

$$u_t = k(C/r)^n \ e^{-\beta_u t} \sin \omega t + k(C/r)^n e^{-\beta_u(t-\Delta\tau)} \sin \omega(t - \omega\tau) +$$

$$+ k(C/r)^n e^{-\beta_u[t-\Delta\tau(m-1)]} \sin \omega[t - \Delta\tau(m-1)] =$$

$$= k(C/r)^n \sum_{i=1}^{m} e^{-\beta_u[t-\Delta\tau(i-1)]} \sin \omega[t - \Delta\tau(i-1)]. \quad \cdots \ (5.11)$$

By putting in formula (5.11) $\Delta\tau = 0$, we obtain (5.10).

For quantitatively assessing the influence of SDB on the resultant displacement velocity when m number of charges collectively act on the compacted soil massif, we shall introduce a coefficient:

$$\eta_t = \frac{u_t}{u_0} = \frac{k(C/r)^n \sum\limits_{i=1}^{m} e^{-\beta_u[t-\Delta\tau(i-1)]} \sin \omega[t - \Delta\tau(i-1)]}{mk(C/r)^n \ e^{-\beta_u t} \sin \omega t}. \quad \cdots \ (5.12)$$

This coefficient indicates how many times lower the seismic effect of SDB is compared to simultaneous blasting.

As we are interested mainly in the maximum displacements, we can rewrite the denominator in (5.12) as

$$mk(C/r)^n e^{-\beta_u t} \sin \omega t \approx mk(C/r)^n \ e^{-\beta_u t}.$$

Then the formula (5.12) becomes

$$\eta_t = \frac{1}{m} \sum_{i=1}^{m} e^{\beta_u \Delta\tau(i-1)} \sin \omega[t - \Delta\tau(i-1)]. \quad \cdots \ (5.13)$$

The summation in this expression is determined as

$$\sum_{i=1}^{m} e^{\beta_u \Delta\tau(i-1)} \sin \omega[t - \Delta\tau(i-1)] = \sum_{i=1}^{m} e^{\beta_u \Delta\tau(i-1)} \times$$

$$\times [\sin \omega t \cos \omega(i-1)\Delta\tau - \cos \omega t \sin \omega(i-1)\Delta\tau]. \quad \cdots \ (5.14)$$

The following relationship is then used:

$$e^{(i-1)\beta_u \Delta\tau} \cos \omega(i-1)\Delta\tau - je^{(i-1)\beta_u \Delta\tau} \sin(i-1)\Delta\tau =$$

$$= e^{(i-1)\beta_u \Delta\tau - j\omega(i-1)\Delta\tau} = e^{(i-1)(\beta_u \Delta\tau - j\omega\Delta\tau)}.$$

Then the sum of series $\sum_{i=1}^{m} e^{(i-1)(\beta_u \Delta \tau - j\omega \Delta \tau)}$ is determined as the sum of geometric progression with the first term being equal to one and denominator $e^{\beta_u \Delta \tau - j\omega \Delta \tau}$:

$$\sum_{i=1}^{m} e^{(i-1)(\beta_u \Delta \tau - j\omega \Delta \tau)} = \frac{1 - e^{m\Delta \tau (\beta_u - j\omega)}}{1 - e^{\Delta \tau (\beta_u - j\omega)}}. \qquad \ldots (5.15)$$

Separating the real and imaginary parts in (5.15) and according to (5.14), we find that

$$\sum_{i=1}^{m} e^{(i-1)\beta_u \Delta \tau} \cos \omega (i-1)\Delta \tau \sin \omega t =$$

$$= \frac{1 - e^{\beta_u \Delta \tau} \cos \omega \Delta \tau - e^{m\beta_u \Delta \tau} \cos \omega m \Delta \tau + e^{(m+1)\beta_u \Delta \tau} \times}{1 - 2e^{\beta_u \Delta \tau} \cos \omega \Delta \tau_+}$$

$$\frac{\times \cos(m-1)\omega \Delta \tau}{+e^{2\beta_u \Delta \tau}} \sin \omega t;$$

$$\sum_{i=1}^{m} e^{(i-1)\beta_u \Delta \tau} \sin \omega (i-1)\Delta \tau \cos \omega t =$$

$$= \frac{e^{\beta_u \Delta \tau} \sin \omega \Delta \tau - e^{m\beta_u \Delta \tau} \sin \omega m \Delta \tau + e^{(m+1)\beta_u \Delta \tau} \sin(m-1)\omega \Delta \tau}{1 - 2e^{\beta_u \Delta \tau} \cos \omega \Delta \tau + e^{2\beta_u \Delta \tau}} \times$$

$$\times \cos \omega \tau.$$

Thus, the coefficient η_t is determined by the formula

$$\eta_t = \frac{1 - e^{\beta_u \Delta \tau} \cos \omega \Delta \tau - e^{m\beta_u \Delta \tau} \cos \omega m \Delta \tau + e^{(m+1)\beta_u \Delta \tau} \cos(m-1)\omega \Delta \tau}{m \left(1 - 2 e^{\beta_u \Delta \tau} \cos \omega \Delta \tau + e^{2\beta_u \Delta \tau}\right)} \sin \omega t -$$

$$- \frac{e^{\beta_u \Delta \tau} \sin \omega \Delta \tau - e^{m\beta_u \Delta \tau} \sin \omega m \Delta \tau + e^{(m+1)\beta_u \Delta \tau} \sin(m-1)\omega \Delta \tau}{m \left(1 - 2e^{\beta_u \Delta \tau} \cos \omega \Delta \tau + e^{2\beta_u \Delta \tau}\right)} \times$$

$$\times \cos \omega \tau. \qquad \ldots (5.16)$$

After trigonometric manipulations expression (5.16) becomes

$$\eta_t = \frac{\sin \omega t - e^{\beta_u \Delta \tau} \sin \omega(t + \Delta \tau) + e^{m\beta_u \Delta \tau} \sin \omega(m\Delta \tau - t) +}{m(1 - 2e^{\beta_u \Delta \tau} \cos \omega \Delta \tau +}$$

$$\frac{+ e^{(m+1)\beta_u \Delta \tau} \sin \omega[t - (m-1)\Delta \tau]}{+ e^{2\beta_u \Delta \tau})}. \qquad \ldots (5.17)$$

Formula (5.17) enables one to calculate the reduction in seismic effect in SDB for different periods (frequencies) of vibrations, the number of groups of delays and delay intervals.

The relationship between the coefficient of seismicity reduction η_t and duration of blast $\tau = (m - 1)\Delta\tau$ at $\Delta\tau = 0.02$ s, $T = 0.2$ s, $\beta_u = 1.8$ is shown in Fig 53. As can be seen from the curve, the change in coefficient η_t based on the duration of blast is periodic in nature. The envelope shown by a continuous line in Fig. 53 reflects the overall nature of the relationship between η_t and τ, which accords with experimental data and the work of other researchers. The envelope for blasting conditions encountered in loess soils is described by

$$\eta_t = e^{-0.0032m\Delta\tau}. \qquad \ldots (5.18)$$

This relationship (5.18) can be used in engineering calculations for determining the degree of reduction in seismic effect in case SDB is adopted.

Screens for Compacting Collapsible Soils by Means of Blasting: SDB does not always give the desired results, i.e., obtaining the permissible velocity of vibrations at foundations of structures. In such cases screens offering seismic protection are used, made in the shape of holes or a series of pits filled up with porous material or soil.

The effect of screening is quite adequately described in sections 5.1 and 5.2. To determine the coefficient of screening of seismic waves, formula (5.6) is applied. Using this formula coefficients for various types of soils have been calculated (Table 15).

The Research Institute for Building Materials of Gosstroi, USSR adopted blasting technology for compacting collapsible loamy soil by using this method for controlling the seismic effects in Groznyi city for preparing the foundation of a children's park.

Such methods have also been adopted for compacting collapsible soils in the construction of a central water supply system in the northern regions of Turkmenia (in Kazandzhik City) by the Branch of Blasting Geodynamics attached to the Geophysics Institute of Academy of Sciences, Ukrainian SSR and

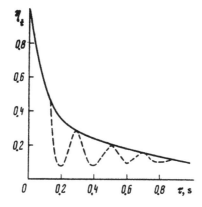

Fig. 53: Relationship between η_t and duration of blast τ.

Table 15

η	f, Hz at $v_2 = 100$ m/s $d_w = 0.5$ m	η	f, Hz at $v_2 = 100$ m/s $d_w = 1$ m	η	f, Hz at $v_2 = 200$ m/s $d_w = 0.5$ m	η	f, Hz at $v_2 = 200$ m/s $d_w = 1$ m
			Loamy sand ($v_1 = 300$ m/s)				
0.98	2	0.955	2	0.988	2	0.993	2
0.955	4	0.923	4	0.984	4	0.974	4
0.915	6	0.682	6	0.976	6	0.943	6
0.837	8	0.52	8	0.974	8	0.858	8
0.764	10	0.365	10	0.97	10	0.847	10
			Loam ($v_1 = 500$ m/s)				
0.826	2	0.901	2	0.996	2	0.99	2
0.81	4	0.7	4	0.99	4	0.964	4
0.8	6	0.482	6	0.979	6	0.922	6
0.7	8	0.325	8	0.964	8	0.867	8
0.58	10	0.215	10	0.945	10	0.802	10
			Clay ($v_1 = 700$ m/s)				
0.951	2	0.831	2	0.995	2	0.985	2
0.831	4	0.542	4	0.984	4	0.944	4
0.682	6	0.331	6	0.967	6	0.883	6
0.542	8	0.203	8	0.944	8	0.803	8
0.424	10	0.123	10	0.915	10	0.718	10

the Institute for Seismic-resistant Designs, Turkmenian SSR [31]. Adoption of such methods was necessitated by the fact that in an operating water supply system, the groundwater reservoirs, pumping station and several other structures are situated at a distance of 100–200 m from each other. An SDB system, using screens consisting of pits and of partitions up to a depth of 4 m and 1.2 m width, was chosen. The pits were dug by an EO-3322A shovel. As a result of these measures, the coefficient of seismicity reduction calculated by formulae (5.7) and (5.18) was 0.18–0.2, which facilitated in lowering the velocity of seismic vibrations to 6–8 cm/s, i.e., the seismic effects were reduced 5–6 times.

As a second example of compaction of collapsible soils by blasting we can consider the Burlinskii irrigation system (Altai region). To compact these soils a blasting scheme using trench charges was adopted. This scheme considerably reduced the time for preparing foundations under structures and consequently reduced the total time for commissioning the buildings and structures into service.

Operations involving compaction of collapsible soils were carried out in confined field conditions, characterised by proximity to residential houses, a grain-drying plant and an underground communication cable. In order to protect neighbouring structures from seismic effects induced by blasting, special measures were developed. These included:

1. Formation of screening cavities (machine-shovelled) alongside the construction area, starting from the direction of the protected objects.
2. SDB of trench charges in which the detonation process developed in the direction opposite to the protected objects.

Screening pits were shovelled up to a depth of 4–5 m and width of 1–1.2 m.

The calculations relating to reduction in seismic effects due to the adoption of these measures are given below. We have denoted the coefficient of reduction in intensity of seismic vibrations as k_r. It is a quantity showing the ratio of soil particle velocities in the wave after (u_3) and before (u_1) passing through the cavity:

$$k_r = u_3/u_1 \text{ or } k_r = 0.2(h/\lambda)^{-0.22},$$

where h is the width of cavity, m.

The length of the seismic wave was determined according to the formula $\lambda = v_p T$ and the wave period by $T = k_s C^{0.145} r^{0.05}$,

where k_s is the coefficient that considers soil conditions, depending on moisture (for loess-type pre-moistened soils, $k_s = 0.055$);

r is the distance from blast site to the protected object, m.

The value of T ranged from 0.18 to 0.22 s for soils in a construction area of an industrial base. When screens using cavity-partition were used in the given conditions, the screening coefficient for seismic waves, took the values $k_r = 0.65 - 0.7$.

The level of reduced seismic effect with SDB was assessed by the formula

$$k_t = \exp(-0.0032 \, n_c \Delta\tau),$$

where n_c is the number of group of charges;

$\Delta\tau$ is the delay intervals, ms.

The calculated values of k_t depending on n_c and $\Delta\tau$ are given below:

k_t	0.73	0.77	0.83	0.79
n_c	5	4	4	5
$\Delta\tau$, ms	20	20	15	15

The general coefficient of reduction in seismic effects, as a result of the above measures, is given by

$$k_{gr} = k_r k_t.$$

The displacement velocity u_1 of the seismic wave was also determined prior to implementing protective measures.

In the given case, trench type cylindrical charges of 100–150 m length, dispersed length-wise in the site, were used. To determine the displacement velocity (cm/s) in the near-field zone of blast effect in a direction perpendicular to the charge axis, the formula given below was used

$$u_p = k(H/\sqrt{C_1})^{0.9}(r/\sqrt{C_1})^{-1.6},$$

while for the far-field zone, the relation used was

$$u_R = kH_0^{0.3}(r/C_{ef}^{1/3})^{-1.3},$$

where k is the coefficient dependent on soil properties (for loess-type soils $k = 300$); depth of trench $H = 4$ m; $C_1 = 10$ kg/m; and $H_0 = H/C_1^{1/3}$ is the scaled depth of placement of trench charge.

The following objects were situated within the zone of seismic effects of the blast: an underground communication cable at a distance of $r = 70$ m from the blast site; a grain-drying plant at a distance of $r = 300$ m and residential houses at a distance of 55 m. Suggested values of velocities in the absence of any protective measures would be: at the cable site $v_{cab} = 2.44$ cm/s; at the grain drying plant $v_{gr} = 0.35$; at the residential houses $v_{rh} = 2.8$ cm/s.

Among all the protected objects, the residential houses proved the least stable from a seismic point of view. The permissible velocity of soil particles at the foundation for such houses would be 1 cm/s, i.e., corresponding to 4 points in the intensity scale of blast-induced seismic vibrations. The suggested intensity of seismic effect induced by blasting without protective measures corresponds to 5 points. Therefore, adoption of protective measures was necessary for residential houses.

The use of SDB enables a reduction in velocity of soil particles up to 2.04 cm/s in residential areas to be achieved.

To bring the velocity of soil particles to permissible levels for residential areas (1 cm/s), using screening cavities, simple calculations showed that the coefficient of screening should be equal to 0.49. With a single cavity the value of k_r is 0.7. Hence it becomes necessary to create a double row of screening cavity which is usually done when cavities are separated by 5 m.

Let us calculate the dynamic load on the underground communication cable as well since it is also an important object. We make use of formulae (P.Z. Lugova et al., 1981) for this purpose:

$$P_T = 3.1 \times 10^5 (r/\sqrt{q})^{-1.15};$$

$$P_\perp = 28.5 \times 10^5 (r/\sqrt{q})^{-0.8},$$

where P_T, $P_\perp$ is the corresponding pressure at the butt ends of trench charges and in a direction perpendicular to the axis of trench charges, Pa:

q is the explosive energy per unit length of charge ($q = GC_1$, where G = heat generated in the blast). $q = 42288.7$ kJ/m.

Should no protective measures be taken, pressures on the cable at a distance of 70 m would be:

$$P_T = 10.7 \times 10^5 \text{ Pa} \;;$$

$$P_\perp = 67.7 \times 10^5 \text{ Pa}.$$

Dynamic load P_{dl} on the cable, considering the effect of screening cavity

$$P_{dl} = 4A_1 A_2 P_{in}/(A_1 + A_2)^2,$$

where A_1, A_2 — acoustic rigidities of water-saturated loess type soil and air (filler in the screening cavity); $A_1 = 3.1 \times 10^6 \text{ kg/cm}^2 \cdot \text{ s}$; $A_2 = 425 \text{ kg/m}^2$;

P_{in} — input (incident) pressure on the screen (at $r = 10$ m for screen site); $P_{inT} = 100 \times 10^5$ Pa; $P_{in\perp} = 925 \times 10^5$ Pa.

Substituting this data in the formula for determining P_{dl}, we obtain the dynamic load on the cable, taking into account the screening effect:

$$P_{dlT} = 0.05 \times 10^5 \text{ Pa};$$

$$P_{dl\perp} = 0.51 \times 10^5 \text{ Pa}.$$

The dynamic load thus obtained is considerably lower than the minimum permissible load on the cable, which was confirmed by results of the blast.

For compacting collapsible loess soils the blasting method has proven to be highly productive and effective. In spite of the fact that preparatory blasting operations were carried out in complex seismic conditions, the protective measures undertaken ensured safety of blasting operations as well as the required quality of soil compaction and protection of surrounding objects.

Using the blasting method, the time for preparation of foundations was reduced by 0.6 year compared to the base variant, i.e., method of wetting and ramming (compacting), and furthermore capital assets could be brought into operation earlier. The economic effect due to the introduction of this method was 132,700 roubles.

Seismic Effects Induced by Drill Hole Charges of Various Configurations Used in Hard Rock Blasting: Apart from other measures, it is possible to achieve target-oriented distribution of energy of seismic vibrations in blasts by using various configurations of charges. With the help of drill hole charges, it becomes possible to widely vary the constructional elements of a charge — subgrade and stemming, number of different parts in a charge and the type of explosives in each, location of primer, number of intervening spaces etc. The parameters of the blast pulse are changed by a combination of the above elements, thereby influencing the nature of propagation of seismic waves. Currently, no recommendations are available for seismically assessing charges of various constructions. Investigations were therefore conducted on this aspect under mine conditions [10].

Seismic receivers of the type VBP-III and VEGIK coupled with the oscillograph N-700 and galvanometers M002 and M001.2 were used for recording
the velocities of ground movement. Parameters of drill hole charges remained
constant in all the experiments except for differences in their configurations.

Continuous charges using one type of explosive (ammonite 6 ZhV and
grammonite 60/70) in field blasts were compared with the suggested continuous
charges in combination with those dispersed by various spacers. The suggested
charges were based on igdanite, selected for introduction in the particular open
pit mine. Single drill hole charges were blasted in an open pit bench of 13 m.

It can be seen from Fig. 54 that the maximum displacements of soil particles occurred when the charge consisted of brisant explosive ammonite 6 ZhV.
Displacement a^z in this case within the near-field of blast effect was 1.5–2 times
more compared to the charge containing grammonite 30/70, while displacements
are almost equal in the far-field zone.

Charges containing igdanite and ammonite and grammonite occupied an
intermediate position according to the intensity of blast-induced seismic waves.
The least soil displacement was observed when charges dispersed with inert layer
(IL) and air gap (AG) were blasted. In this case the value of a was lower by
1.7–2.5 times, compared to the blasting of charges consisting of ammonite and
grammonite. A diametrically opposite picture was noticed in the time-dependent
parameters, but was not so distinctly different as in the case of amplitude of
vibrations. The frequency of ground motion generated by brisant explosives in
the near zone was 10–15% lower than in blasts using igdanite, grammonite or
in the blasting of charges dispersed with IL and AG.

Thus, charges based on igdanite containing IL and AG are the most suitable
for reducing the effect of seismic vibrations. Moreover, charges with ILs are

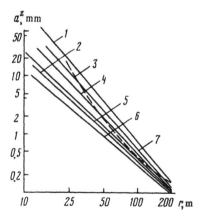

Fig. 54: Relationship between displacement velocities of massif and epicentral distances and construction of charge (1–7) correspond to the serial number of charge according to Table 15(a).

Table 15(a)

No. of drill hole charge	1	2	3	4	5	6	7
Weight, kg:							
ammonite 6 ZhV	126	—	63	—	21	21	42
grammonite 30/70	—	126	63	42	42	42	—
igdanite	—	—	—	100	50	50	100
Height, m:							
inert filler	—	—	—	—	1	—	—
air gap	—	—	—	—	—	1	—
water layer	—	—	—	—	—	—	1
Depth of drill hole, m	13	13	13	13	13	13	13
Stemming length, m	6	6	6	6	6	6	6
Diameter of drill hole, mm	150	150	150	150	150	150	150
Subdrilling, m	1.5	1.5	1.5	1.5	1.5	1.5	1.5
Line of least resistance, m	4.8	4.8	4.8	4.8	4.8	4.8	4.8

technologically more acceptable as very little time is required to form such layers in the drill holes. Therefore, such charges were recommended for blasting operations in a limestone quarry, Karpenyai, Latvian SSR. The plant area and cement factory are located at a distance of 200–230 m from the quarry.

It is to be noted that in this quarry single-row blasting was adopted. With the introduction of charges of the above configurations, particularly in seismic-unsafe sections, measurements were taken both for single-row and multirow blasting with the object of switching over to the latter type of blasting. In the former case, the charges used were continuous while in the latter, they were dispersed with IL. Twelve large-scale blasts were conducted. It was established from these experiments that in multirow blasting of charges dispersed with IL, the buildings are not affected from the point of view of vibration. The recorded maximum vibrations (displacement velocities) for the horizontal component (which is the most harmful for structures) were only 0.3–0.35 cm/s while the permissible velocity is 3 cm/s (Table 16). An analysis of the data showed that adoption of multirow SDB ensures safety of operations and reduces further the seismic effects of blasting on the protected objects, the scale of blast notwithstanding (Fig. 55).

Introducing charges of seismic-safe configurations, based on the results of investigations conducted, not only enabled exploitation of mineral even in the near-field zone, but also permitted a changeover from two-bench mining (each bench of 11–13 m height) to single-bench mining (bench height 18–21 m) as well as a change in the advance of the working front to a rational direction. The possibility of the latter aspect was realised by introducing multirow SDB as it utilises more completely the optimisation conditions of blasting methods.

Table 16

No. of blasts	Total weight of charges, kg	Place of installation of sensors	Distance from obseva- tion point, m	Component	Period of of vibra- tions, s	Displacement velocity along component, cm/s
3	9240	Pipe shop:				
		soil	1560	x	0.11	0.081
				y	0.1	0.071
				z	0.1	0.155
		IV Floor	1560	x	0.45	0.232
				y	0.39	0.15
				z	0.14	0.225
4	11792	Pipe shop:				
		soil	1600	x	0.16	0.148
				y	0.15	0.086
				z	0.16	0.216
		IV Floor	1600	x	0.38	0.39
				y	0.36	0.192
				z	0.12	0.35

Note: No. of group charges 7; delay interval 70 ms.

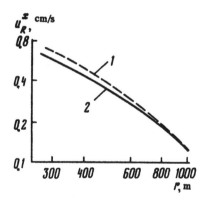

Fig. 55: Dependence of massif displacement velocities u_R^z on the epicentral distances r in (1) single-row blasting and (2) multirow blasting.

As a result of these studies, the weight of explosives that would ensure seismic safety for a particular group of delay intervals was established for the Karpenyai quarry. The weight of explosives for the southern flank of the quarry was not to exceed 500 kg and that for the northern flank 1500 kg.

Influence of Characteristics of Slope and Working Flank on Seismic Effects Induced by Blasting (written in collaboration with S.N. Markelov): A series of studies conducted at home and abroad provide data on the influence of slope on the seismic effects of large-scale blasts. In particular, an increase in the intensity of vibrations at bench edges was noticed with the velocity of vibrations increasing considerably near the bench edge compared with that on the opposite part of the bench. In the central part, the changes were insignificant [27]. It was also observed that, independent of the level of block in which blasting is to be done, the velocity curve ascends approximately at the same distance from the blasting site in each level. Minimum displacement velocities were observed at 80–100 m (for the conditions described) from the blast independent of the location of the blasting block, which speaks of the non-uniformity in wave propagation along the pit flank; at a specific distance from the blast site there is a tendency for the velocities to increase. Formulae are given for calculating the intensity of seismic waves considering the mutual elevation of blasting blocks relative to the protected objects [9].

In neither the above-mentioned research nor literature published earlier was the nature of the wavefield during the propagation of seismic vibrations considered in its entirety while studying and forecasting the seismic effects of blast. As is well known, a surface wave is very harmful to structures as it is characterised by large amplitude, weak attenuation over distance and low frequency of vibrations compared to body waves. A surface wave is formed at a certain distance from the source of seismic vibrations, which depends on properties of rocks and depth of charge placement (approximately equal to 5 times the depth). However, if on the path of a surface wave an obstruction is placed, for example a bench, then the normal propagation is broken and on the bench surface such a wave is again formed owing to the mutual action of body waves. Thus a zone is formed behind the bench in which body waves of different types predominate, such as reflected, refracted and deformed, while lower part of the bench plays the role of an 'imaginary' source of vibration. In this case, the distance at which the surface wave begins to form is determined not only by the depth of charge placement and level difference between the blasting block and the protected object, but also by the height of bench located in the path of wave propagation.

The seismic effects induced by trial and field blasts and the wavefield formed during the blasts at the open pit of Skala-Podol'sk asphalt concrete factory were studied to establish seismic parameters for safe conduction of blasting operations.

One of the main factors influencing the seismic effect was the presence of benches (8 and 20 m in height) between the blast site and the protected objects. Benches were composed of clay with interlayers of sandstone (Fig. 56). The distance between the edge of the upper bench and the protected object was 90 m.

The following data were obtained from an analysis of seismograms.

Initially vibrations with a frequency of 25–30 Hz were distinctly visible on the seismogram. The vertical component of these vibrations exceeded by

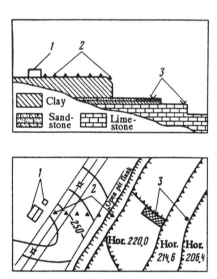

Fig. 56: Schematic section of western flank of open pit of the Skala-Podol'sk Factory of Asphalt-Concrete:

1 — industrial buildings of the factory; 2 — locations of seismic receivers; 3 — blast site.

6–8 times the horizontal component. These vibrations were followed by vibrations of 8–13 Hz frequency for which a change in the ratio of intensity of vertical and horizontal component was characteristic as they moved away from the pit flank: if at a point nearer to the pit flank the vertical component exceeded twice the horizontal, then at a 20 m distance the amplitudes were nearly equal, while at 60–80 m, the horizontal component exceeded the vertical by 8–10 times; farther away from the pit flank, the intensity of vibrations gradually evened out.

Further, an increase in the phase difference of maximum values of vertical and horizontal components of seismic waves was noticed.

Within the range of distances under investigation the displacement velocity varied between 0.1–0.8 cm/s and the maximum value was attained by the horizontal component at 60 m distance (Fig. 57).

An analysis of the wavefield enabled the conclusion that the measurements were taken in a zone in which surface waves just begin to form while body waves play a significant role as they propagate inside the rock massif. The maximum displacement velocity of seismic vibrations is not observed at the bench edge, but at a distance of 40–60 m from it, which corresponds to the formation zone of the surface wave. Some screening effect of the benches is noticed, which is expressed by means of reduced intensity of seismic vibrations near the bench edge, predominance of vertical component of seismic vibrations in this zone and displacement of the maximum of vibrations from the bench edge (Fig. 57).

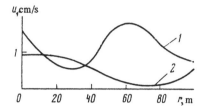

Fig. 57: Relationship between displacement velocity u and distances up to edge of top bench: 1 — x-component; 2 — z-component.

Based on the results of investigations the following conclusions may be drawn:

— Surface waves pose the greatest hazard for buildings due to their larger amplitude compared to that of body waves, a marked presence of low frequency vibrations and lower attenuation over distance.

— The ratio between the height and number of benches on the pit flank and distances up to the protected object greatly affect the formation of a surface wave and consequently the seismic effect in conditions of complicated relief at the pit flank, compared to the influence exerted by depth of block being blasted or difference in elevations of blast site and observation point.

— A peak in seismic vibrations is noticed at a certain distance from the bench edge; in the presence of benches the seismic effects of a blast are reduced due to the screening of the 'initial' surface wave, which makes it possible to reduce the harmful effect of a surface wave on objects located near the pit flank through creation of high benches along the flank.

— While assessing the seismic hazard due to large-scale blasts, it is necessary to consider the surface relief at the pit flank between the block being blasted and the protected object. The prediction of seismic effects by means of currently available analytical relationships is less effective in the presence of benches in a zone of complex relief. Instrumented measurements in each specific case are required in order to establish the level of seismic vibrations induced by blasting.

Mutual Relationship between Seismic and Crushing Effects Due to the SDB of Charges (written in collaboration with S.N. Markelov and Yu.A. Sikorskii): The usual recommendations about the influence of delay interval on the results of blasting operations do not always successfully solve the engineering problems encountered in various geo-mining conditions. Experiments were conducted in limestone and granite quarries of Ukraine to obtain a uniformly crushed mined mass with a given granulometric composition as well as to ensure minimal seismic effect, especially in congested mining conditions. Drill hole charges with different delay intervals were blasted (Fig. 58). The effect of blasting these charges was studied on the basis of known methodologies and assessed by the value of displacement velocity u, volume of crushing cone Q_d and yield of large fragments δ_K and small fragments δ_M of the mined mass contained in the volume of the crushing zone. The functiona! relationships of $u, \frac{I}{2}, Q_d, \delta_K$ and δ_M and delay interval $\Delta\tau$ (Fig. 59) were derived by processing the experimental data.

Analysis of the figure shows that the peak value of u is observed for vertical components of ground movements and the lowest value for horizontal. Reduction

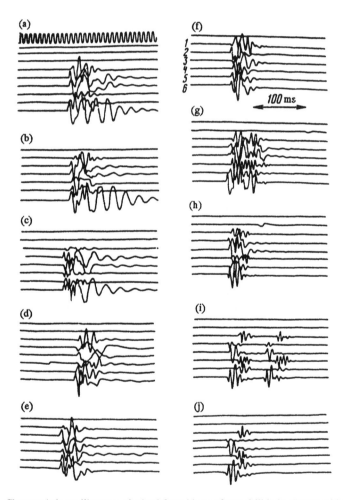

Fig. 58: Characteristic oscillograms obtained from blasts of two drill hole charges with delays: a — 10 ms; b — 30; c — 45; d — 75; e — 105; f — 120; g — 150; h — instantaneous blasting; i — strapped charges with 105 ms delay; j — single strapped charge; 1, 2 — horizontal component y_2 and y_1 respectively; 3, 4 — horizontal components x_1 and x_2 respectively; 5, 6 — vertical components z_2 and z_1 respectively.

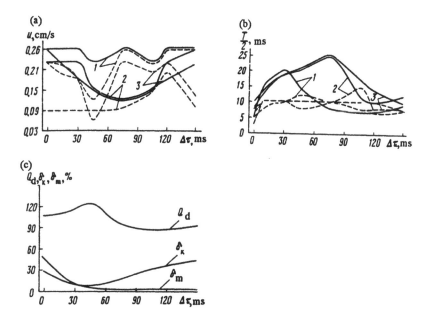

Fig. 59: Change in (a) displacement velocities, (b) half cycles of vibrations at distances 5 m (continuous curves) and 15 m (dashed curves) depending on the delay intervals:

1 — vertical z-component; 2, 3 — horizontal components x and y respectively; c — variation in indices of fragmentation depending on the delay interval.

in the intensity of vibrations is noticed in the 30–80 ms interval, and the maximum reduction is established at $\Delta\tau = 45$ ms. At $\Delta\tau$ exceeding 80–100 ms, the level of vibrations equals to its value obtained during instantaneous blasting.

The empirical relationships of variation in values of u along horizontal components x and y depending on $\Delta\tau$ can be satisfactorily approximated by the following formulae:

$$u_x = 0.25 \times 10^{-4}(\Delta\tau - 75)^2 + 0.12;$$

$$u_y = 0.2 \times 10^{-4}(\Delta\tau - 75)^2 + 0.12.$$

The conclusions conform to the data on charges in values on semi-periods of vibrations.

The analysis of rock fragmentation also shows that the best indices are obtained at $\Delta\tau = 25$–40 ms. In this case the volume of crushed rock is higher by 17–23%, yield of coarse fractions is lower by 35–45% and the fine fractions more by 22–28% [7].

Thus the adoption of rational delay interval for mined rocks and blasting schemes ensures a reduction in seismic effect of blast and simultaneously improves the quality of fragmentation of the mined mass.

The conclusions drawn from the results of these investigations were taken as the basis for framing recommendations for the safe conduct of blasting operations in the near-field zone (up to 70 m) in the open pit belonging to Skala-Podol'sk Asphalt Concrete Factory. It was recommended that among the series of drill hole charges, the first two or three be blasted after 30–35 ms and the remaining after 45 ms.

Introduction of the above recommendations in the given geo-mining conditions allowed the exploitation of additional reserves in the zone up to 200 m from the industrial plant without stopping the plant operations and without incurring any capital expenditure on repair and maintenance of industrial structures and buildings.

Natural Seismoanisotropy (written in collaboration with V.V. Zakharov): Seismoanisotropy is defined as the non-uniform response of rock massifs to the propagation of seismic waves in various directions from a blast source. Here it should be noted that the rock massif *per se* is not disturbed by any external effect and exists in its natural state. Consequently, while blasting in hard rocks the protected objects located at the same distance from the blast site are subjected to different degrees of seismic effects.

As a result of investigations conducted in granite and limestone quarries, the variation patterns of seismic wave characteristics depending on the structure of such rock massifs were established [3, 4]. Further recommendations were given for seismic-safe conduct of blasting operations [3, 5–7], which were subsequently refined by the conclusions drawn on the influence of blast-induced seismic effects in hard rocks covered by overburden. These prescriptions were the basis for supplementary measures for determining seismic-safe parameters [30].

In mines the thickness of overburden may vary over a wide range. Therefore, in such geo-mining conditions the blasting operations are to be conducted based on analytical investigations and experimentation to safeguard mining equipment and other objects situated nearby.

Let us analyse the case of blasting operations in overburden consisting of sedimentary rocks on a hard rock base in which seismic measurements were taken. In the first approximation, the problem reduces to a consideration of vibrations at the free face of the layer lying on a rigid semi-space. Seismic energy is transferred from the source of vibrations over the free face of semi-space by a Rayleigh surface wave. If an overburden does not exist, then the complex vibrations contained in this wave propagate without distortions.

The presence of an overburden layer on hard rock base leads to selective strengthening of vibrations in a wave of a particular length depending on the ratio of layer thickness and wavelength. E.F. Sakhanov mentions that the amplitude of vibrations at the surface of a layer lying on a rigid semi-space can be determined by the formula

$$A = A_0 \frac{2}{\sqrt{\cos^2 \alpha_0 + \overline{m}^2 \sin^2 \alpha_0}} e^{j\overline{\varphi}}, \qquad \dots (5.19)$$

where A_0 is the wave amplitude at a rocky base;

α_0 is $2\pi h_c/\lambda_w$;

$\overline{m}$ is $\rho_1 v_{p_1}/\rho_2 v_{p_2}$;

h_c is the thickness of overburden layer;

λ_w is wavelength;

ρ_1, ρ_2 and v_{p_1}, v_{p_2} are density and propagation velocity of a longitudinal wave within the boundaries of the overburden layer and bedding semi-space respectively.

The quantity $2\left/\sqrt{\cos^2 \alpha_0 + \overline{m}^2 \sin^2 \alpha_0}\right.$ in (5.19) determines only the amplitude of vibrations at the free face of the layer, while $e^{j\overline{\varphi}}$ considers its absorption properties. Evidently, in the expression for determining only amplitude of vibrations, at a value α_0 corresponding to the ratio $h_c/\lambda_w = \overline{n}(\overline{n} = 0, 1, 2, \ldots, k)$, the amplitude is quantified only by $\cos^2 \alpha_0$ and conforms to a twofold increase in amplitude at the free face. At α_0 equal to the ratio $h_c/\lambda_w = 0.25 + \overline{n}$, the amplitude of vibrations is determined entirely by the value of coefficient $\overline{m}$ or by the ratio of acoustic rigidities of the layer and semi-space.

For clarity, let us assume that the semi-space is formed by granitic gneiss or limestones, while the overburden consists of clays having the following parameters:

Rock	Clay	Limestone	Granitic gneiss
v_p, km/s	0.6	3.4	6.4
ρ, t/m^3	1.8	2.4	2.7

The calculated value of amplitude indicates that in this case the coefficient $\overline{m}$ for a rocky base in the form of limestones takes the value $\overline{m}_1 = 0.13$, while for granites $\overline{m}_2 = 0.06$. The maximum enhancement in amplitude corresponding to the ratio $\alpha_0 = \dfrac{\pi}{2} + \overline{n}\pi$ or $h_c/\lambda_w = 0.25 + \overline{n}/2$, is 7.7 A_0 for limestone and 16.12 A_0 for granites.

The general form of relationship between amplitudes and layer thickness depending on the absorption capacity is shown in Fig. 60,a, when the coefficient of absorption $\overline{\varphi} = -0.5 \ m^{-1}$.

Thus the loose layer lying on a rocky foundation can significantly amplify the amplitude of vibrations at the free face. However, absorption is considerably more in this layer than in the rocky base, and in each specific case it is necessary to obtain the essential parameters with the help of instruments. This condition must be satisfied since the characteristics of the overburden soil layer influence the variation in frequency spectrum of seismic waves.

From the aforesaid, it follows that the layer is essentially a narrow strip-like filter that enables particular frequencies to pass through. Considering that

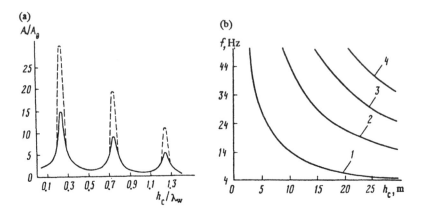

Fig. 60: Relationship between (a) vibration amplitudes A/A_0 and ratio of layer thickness to wavelength and (b) resonant frequency and layer thickness:

$1-n = 0$; $2-n = 1$; $3-n = 2$; $4-n = 3$.

$\lambda_w = v_p T$, the prevalent frequencies in vibrations can be predicted and the degree of hazard due to seismic effects can be assessed in advance.

The velocity of particles in the soil surface is taken as the criterion for evaluating danger due to seismic effects. To determine the values of velocity, we take the derivative from (5.19). Then

$$u = u_0 \kappa_u e^{-\bar{\sigma} h_c}, \qquad \dots (5.20)$$

where $\kappa_u = 4\pi f \Big/ \sqrt{\cos^2 \alpha_0 + \bar{m}^2 \sin^2 \alpha_0}$;

$\bar{\sigma}$ = coefficient of absorption.

κ_u assumes maximum values at $\bar{m} < 1$, which conforms to parameters of the layer made up of sedimentary rocks and to values of α_0 which satisfy the condition $\alpha_0 = \pi/2 + \bar{n}\pi$ or $h_c/\lambda_w = 0.25 + \bar{n}/2$ when the ratio of the frequency to thickness of layer $f = (v_p/h_c)(0.25 + \bar{n}/2)$.

Thus the frequencies satisfying the condition $f = (v_p/h_c)(0.25 + \bar{n}/2)$ would correspond to the maximum values of velocity observed. In Fig. 60,b, the resonant harmonic frequencies are given for the most harmful part of the spectrum 5–50 Hz depending on the layer thickness. Curve 1 corresponds to wavelengths $\lambda_w/4$, curve $2 - \lambda_w/2$, curve $3 - 3/4 \lambda_w$ curve $4 - \lambda_w$.

The presence of the frequency term in the numerator in the expression for κ_u does not indicate an increase of κ_u to infinity as $f \to \infty$, since with an increase in frequency the absorption of high frequency components in the spectrum also increases in sedimentary rocks. Any decrease in the value of u due to absorption is much greater than due to an increase in frequency.

If we limit ourselves to the relationship $f = F(h_c)$ at $\bar{n} = 0$, then (5.20) is written

$$u = u_0 \frac{\pi v_p}{h_c \sqrt{\cos^2 \alpha_0 + \bar{m}^2 \sin^2 \alpha_0}} \, e^{-\bar{\sigma} h_c}.$$

As already mentioned, in anisotropic massifs the change in velocities is associated with the direction of wave propagation. The wavefront of blast-induced seismic vibrations can be approximated with a high level of confidence to that of an ellipse. In the ellipse, the magnitudes of minor and major axes are determined by displacement velocities in mutually perpendicular directions. The major axis coincides in direction with the strike azimuth of the principal fracture system. Then, taking the expression for functions of variation in velocities at the rocky foundation with distance $u_{II} = k_{II} \left(C^{1/3}/r \right)^{n_{II}}$ and $u_\perp = k_\perp \left(C^{1/3}/r \right)^{n_\perp}$ respectively as half of minor and major axes of ellipse and writing the expression for u_0 in the polar system of co-ordinates, we obtain

$$u_0 = \frac{u_\perp^2}{u_{II} + \sqrt{u_{II}^2 - u_\perp^2} \, \cos \varphi_a},$$

where φ_a is the angle between strike azimuth of the principal fracture system and the given direction.

Displacement velocity in any direction from the blast site, considering the thickness of the overburden layer, is given by

$$u_r = \frac{u_{II}^2}{u_\perp + \cos \varphi_a \sqrt{u_{II}^2 - u_\perp^2}} \frac{\pi v_p}{h_c \sqrt{\cos^2 \alpha_0 + m^2 \sin^2 \alpha_0}} \cdot e^{-\bar{\sigma} h_c}. \quad \dots \, (5.21)$$

Seismic microzoning, taking into account the thickness of overlying strata, can be done with permissible accuracy by using (5.21).

Thus consideration of the natural seismoanisotropy of a massif in blasts carried out in hard rocks with an overburden of soils requires a special approach for determining safe blasting parameters for protecting industrial and civilian structures. Studies conducted in limestone and granite quarries have established patterns in the variation of velocity of seismic vibrations due to natural and technological factors. In the course of such studies, it was confirmed that the velocity contour of soil particles in a blast from a single charge is elliptical in shape with the ratio of major to minor axis equal to 1.1–1.4 and sometimes even more. The established relationship on the variability of the magnitude of seismic vibrations at equal distances from the blast site is determined, apart from other factors, by the influence of anisotropy of joints of rock massifs. Peak velocity is observed in the direction of the major axis of the ellipse (zone of crushing), corresponding to the strike azimuth of the principal system of vertical joints in the rock mass.

Distance from blast site, m	100	100	200	200	100	100	200	200
Strike azimuth of joints, degrees	170	350	170	350	260	80	260	80
Averaged values of u at points of measurement along mutually perpendicular direction, cm/s	$\dfrac{1.3}{4.9}$	$\dfrac{1.8}{5.4}$	$\dfrac{0.9}{2.8}$	$\dfrac{1.2}{3.8}$	$\dfrac{1.1}{2.6}$	$\dfrac{1.0}{2.3}$	$\dfrac{0.7}{1.8}$	$\dfrac{0.7}{1.3}$

Note: Values of u for limestone are given in the numerator, those for granite in the denominator.

To substantiate the seismic-safe blasting parameters established by experimental data, it is necessary to know the change in values of coefficients k and n found in the well-known formula of M.A. Sadovskii. These coefficients are determined by solving a system of two equations with two unknowns (k and n). They characterise the displacement velocities u_1 and u_2 at different distances r_1 and r_2 from the blast site in a given direction [4][1].

After simple transformations, the displacement velocities in the direction of the principal joint system u_{II} and perpendicular to it $u_{\perp}$, are determined. Subsequently, substituting u_{per} in the expression for obtaining displacement velocities isolines of equal velocities of seismic vibrations can be obtained. In this case equal values of u_{per} would be observed from the blast site, along the two directions, at distances r_{II} and $r_{\perp}$ ($r_{II} = r_1$ and $r_{\perp} = r_2$).

Taking r_{II} and $r_{\perp}$ as major and minor axes of ellipse of a seismic-safe zone, after simple transformations the equation for obtaining seismic-safe distance in any azimuthal direction in hard rock blasting and the protected object being located on such a foundation, can be written in the following form:

$$r_{c_1} = \frac{0.87 C_2^{1/3} (k_1 u_{per}^{-1})^{1/n_1}}{\sqrt{1 + \left[k_2^{n_2}/k_1^{n_1}(u_{per})^{1/n_2 - 1/n_1} - 1\right] \sin^2 \varphi / \sin^2 \alpha}}, \qquad \dots (5.22)$$

where C_2 is the charge weight for group delay, kg;

 k and n are coefficients that depend on rock properties and indices of blast effect respectively;

  φ is the angle between azimuth of the principal joint system and direction of propagation of seismic vibrations towards the object, degrees;

  α is the angle between experimental profiles, degrees.

Using formula (5.22) in calculations, we can vary the main blasting parameters in case the principal direction of propagation of seismic vibrations is

---

[1] Vorob'ev was the first to suggest that the radius of a seismic-safe zone could be determined by an elliptical equation.

oriented within 0–360°. In order to determine k and n, it is necessary to conduct experiments along two profiles arbitrarily situated to each other.

The existing relationships were obtained based on the results of experimental or large-scale field blasts and are valid only for rock massifs. However, in real conditions the protected objects may be situated either on a rocky foundation or a soft soil foundation. In such a case it is not possible to distinguish the influence of overburden soil covering the rocky foundation while determining seismic-safe parameters for blasting in hard rocks, using well-known calculation methods. Moreover, there are no data on the patterns of variation in seismoanisotropy on the surface of soil when hard-bedded rocks are blasted.

Let us establish the seismic-safe distance taking into account the overburden in hard-bedded rocks (intermediate transformations are dropped);

$$r_{c_2} = \frac{0.87 C_2^{1/3} \left[(u_{so}/B)k^{-1} \right]^{1/n_1}}{\sqrt{1 + \left[(k_2^{n_2}/k_1^{n_1}) (u_{so}/B)^{1/n_2 - 1/n_1} \right] \sin^2 \varphi / \sin^2 \alpha}}, \qquad \ldots (5.23)$$

where $B = \left[4\pi f v_p \left(h_c \sqrt{\cos^2 \alpha_0 + m^2 \sin^2 \alpha_0} \right) \right] e^{\bar{\sigma} h_c}$, the complex coefficient; u_{so} the displacement velocity of overburden soil.

Using the values of r_1 and r_2, the seismic-safe zone for any charge weight can be established. If the protected object falls in the danger zone, the charge weight should be reduced or while reducing it the screening of blast-induced seismic waves should be done by cavity formation, which happens to be the most effective method.

In the congested conditions of limestone and granite quarries (under the Ministry of Highways, Ukraine), formulae (5.22) and (5.23) are used in recommending measures for conducting seismic-safe blasting operations.

It is necessary to fulfil sequentially the following steps for implementing in practice the method developed:

—Determination of elements of joints in the massif, demarcating particularly the strike azimuth of the principal joint system (or according to regionalised maps of quarry fields).

—Conducting a trial blast to calculate displacement velocities and coefficients k and n along mutually perpendicular directions, one of which should coincide with the direction of the principal joint system.

—Establishing the conditions for locating protected objects, calculating elliptical axes and mapping of ellipse for the chosen charge weight and development of nomograms.

—Mapping onto the pit plan the isolines of seismic-safe zones and the permissible advancing front of blasting operations towards protected objects using the ellipse of the seismic-safe zone for the chosen charge weight.

The isoline of the seismic-safe zone is tangential (enveloping) to the ellipses as well as to the contours of the protected objects (Fig. 61). The line joining

136

geometrical centres of all ellipses will be the line of front of permissible approach of blasting operations, with a fixed charge weight.

It should be noted that the ultimate permissible charge weight in SDB is selected according to the nomogram for a group of charges (one delay interval) according to established rules.

The engineering method of calculating seismic-safe blasting parameters developed herein was introduced in quarries of the Skala-Podol'sk Asbestos Concrete Factory. This method allowed the exploitation of ready-to-mine reserves in the near zone of blast (within 200 m from the protected objects) without incurring any extra expenditure on repair and maintenance of plant buildings in the factory and residential houses of the northern and southern villages. However, for the continued operation of the factory it became essential to mine reserves in a zone within 70 m from the plant. Taking into account the planning tasks, the volume of mineral contained in this zone was expected to be worked out within four years.

The investigations carried out in granite and limestone quarries led to recommendations for the safe conduction of blasting operations. The salient features of these recommendations are:

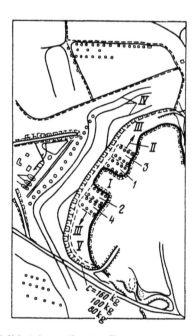

Fig. 61: Plan of Skala-Podol'sk Asbestos Concrete Factory:

I—Opening trench; II—direction of initiation of charges in series; III—direction of extraction of bench; IV—isolines of equal displacement velocities for different charge weights; V—screening cavity; 1–4—sequence of extracting blocks in the bench.

1. Selection of charge for a single delay interval, total charge weight, delay interval between groups of charges and hook-up plans with primer location.

   In certain cases, to ensure seismic safety of blasting operations, benches are divided into two subbenches in some quarries with drivage of opening trench for sequential blasting of blocks in the direction of the working front along both flanks (an example is shown in Fig. 61) .

2. Determining displacement velocity for the given blast parameters based on distances up to protected objects and the permissible values for the given buildings.

3. In certain cases, along the slope contour screening cavities are formed for assured reduction in the seismic effects of blast on the objects as well as to protect the non-working flanks of quarries. For this purpose, the blasting parameters are chosen according to the well-known method of contour (smooth) blasting. The creation of cavities along the contour of isolines of equal velocities of seismic waves ensures screening of their energy compared to linear, linear-discontinuous and parabolic cavities by 2, 1.6 and 1.3 times more respectively.

Artificial Seismoanisotropy (written in collaboration with L.A. Furman): The concept of artificial seismoanisotropy includes the effect of an anthropological factor, i.e., human activity, directed towards changing the capability of rocks to transmit seismic waves, aimed at redistributing the energy carried by them in a given area. The parameters of this area are established by a series of engineering problems whose objective should conform to limiting the intensity of seismic vibrations within permissible norms for the corresponding technological conditions.

Currently, much experience has been gained for reducing the intensity of effects caused by seismic waves. Summarising these results, one can discern that the search for effective solutions has been continued in two directions: (1) reduction in the intensity of effect at the source of vibrations on the rock massif and (2) reduction in the intensity of waves along their path of propagation in the massif.

In our opinion, the perspective of the investigations in the latter direction can be realised with no radical reorganisation of drilling-blasting operations. However, extra expenditure will be incurred for solving problems by this method.

The perspectives in development of engineering methods in the second direction are governed in the ultimate analysis by the growing demand for raw minerals and also by the peculiarities of geo-mining conditions.

Antiseismic screens are created directly near the rock block to be blasted and also near the protected object. In both cases additional expenditure is involved and inadequate knowledge of the problem would tend to nullify the effectiveness of the methods adopted. Moreover, a screening cavity should be created near the rock block before every blast, which complicates the blasting technology.

138

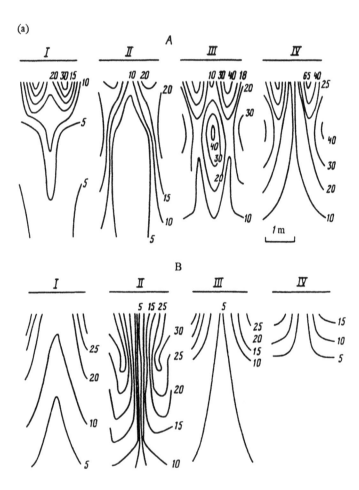

Fig. 62: Distribution of displacement velocities behind (a) a straight screen and (b) semi-circular screen:

A — along the vertical z-component; B — along the horizontal x-component; I–IV maximum values of amplitudes when the time of entry of wave of each corresponding phase is increased.

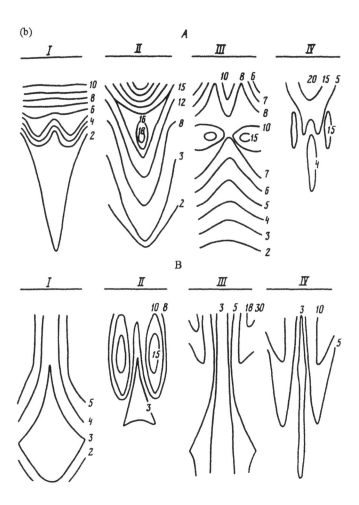

The formation of a cavity near the protected object is more advantageous as the effect of the screening cavity is felt longer and is therefore more economical.

It is to be noted that massifs that have a 'hole type' structure are anisotropic massifs. Cylindrical cavities could be used as 'holes' whose parameters and location enable channelling of the intensity of vibrations in a given direction. This can be achieved by creating several flat screen-cavities. The methods for screening seismic waves are as yet to be fully developed in practice despite recognition of their benefits, as established by results of modelling and trial experiments.

In practice, screening methods are widely adopted in smooth blasting to protect the rock massif behind the designed contour of working. For screening to be effective the cavity dimensions must be commensurate with the propagating wave. Otherwise, the effectiveness of screening would be episodical and that too only for such blasts that generate waves of length not exceeding the selected optimal value.

While investigating the mutual interaction between the wages and the screen, the configuration of the screen chosen was a straight line [17] and only in certain cases was the mutual action between seismic waves and mine workings of cylindrical form [24] considered.

The main parameters that influence the screening coefficient of seismic waves are geometrical dimensions and their ratio with the length of incident wave. In this case the task of enhancing the effectiveness of screening is solved simply by increasing the geometrical dimensions of the screen to bring it nearer to the source of vibrations. The effectiveness of screening is determined largely by the configuration of the screen [17], while at the same time no specific recommendations are given.

It is well known that the amplitude of wave

$$A = A_0 f_1(z_1/z_2) f_2(d_p/\lambda_w) f_3(\varphi) f_4(\sigma_{ab}).$$

where z_1, z_2 are the acoustic rigidities of overburden rocks and screen;

d_p is the width of screen;

λ_w is the length of incident wave;

φ is the angle of incidence of wave on screen;

σ_{ab} is the absorption index of wave in screen.

From the above formula it is evident that by increasing the ratio (z_1/z_2) as well as the angle of incidence φ, we can realistically amplify the effectiveness of screening. As the ultimate value z_2 is indicated for air, then other things remaining equal, for a specific rock massif it has a specific value in the ratio z_1/z_2.

The angle of wave incidence on the screen can be varied by bringing the source of vibrations nearer to the screen and by changing the configuration of the screen itself.

Bringing the source of vibrations nearer to the screen finally leads to increased intensity of seismic waves behind the screen and the need arises for creating the screen itself with geometrical parameters that reduce the level of vibrations to a safe limit. Changing the configuration of a screen is accompanied by the following advantages:

1. Possibility of wave falling on the screen at an angle exceeding the critical.
2. Reduction in intensity of wave behind the screen due to diffraction at edges. This case can be realised very easily. As the wave falls on the screen with a curvilinear configuration, the curvature can be selected in such a way that the edges of the screen are situated farther than the point at which the wave 'slides' into a so-called zone of 'penumbra'.
3. Use of screens that are effective in congested conditions. As the theoretical analysis of wavefields behind screens of different configurations is quite difficult, the authors attempted to compare the wavefields behind a straight screen and behind a semi-circular one in field conditions.

The experimental studies were conducted in hard rock quarries mining different minerals — sandstone, gypsum, limestone and granites. Measurements were taken according to known methodology with the help of seismic receivers SV-30 with ShK-2 complex and oscillograph N-044.3.

Screens were created by blasting with detonating cord. The source of vibrations was a charge of 42 mm diameter with different explosives and a detonating cord varying in number of threads from 1 to 3. The source was situated at 0.5 m intervals from the screen, within the range 0.6–3.6 m.

The fields of displacement velocities behind the screen were drawn according to the results of processed oscillograms. It can be seen from Fig. 62, that as the screen was moved away from the blast source, the velocity reduced: for a straight screen, the calculated values of u varied from 0.6 to 30 cm/s and for a semi-circular screen form 0.6 to 12 cm/s. At the same distance from the source the velocity behind the semi-circular screen was 1.75–2 times less compared to that behind the straight screen. Reduction in values of u behind the semi-circular screen is explained by reduced intensity of diffracted wave in the wavefield, as in this case the wave falls on the screen at an angle nearer to the critical and the edges of the screen are located in the 'penumbra' zone. In the case of a straight screen, this zone is absent and the intensity of diffracted waves is considerably higher.

Thus by creating screens of curvilinear configuration and thereby increasing the degree of artificial anisotropy of the massif, it is possible to regulate objectively the seismic effects of field blasts with maximum effectiveness. The curvature of such screens could be varied (semi-ellipse, semi-circle etc.). The configuration of screens should be selected keeping in view the geo-mining conditions, the state of protected objects and blasting technology.

6

Effectiveness of Protection Measures against Seismic Effects Induced by Blasting[1]

6.1 Methodology for Assessing the Technoeconomic Effectiveness of Protection Measures against Seismic Effects

Perfection of the techniques for undertaking seismic-safe blasting operations is of paramount importance for the national economy. One of the major directions in this task is a comprehensive technoeconomic assessment of perfection of the entire cycle of blasting operations as well as its unit processes and operations and the measures for achieving seismic safety. In this connection a series of problems arise whose solution could lead to more effective extraction of minerals.

From an analysis of normative standards, the results of various investigations and experience gained in technoeconomic evaluation (TE) of blasting results [14, 16, 23], it appears that there is no uniformity of opinion as yet on the overall assessment of the entire blasting cycle as well as of each unit operation. This task is difficult as results of blasting operations are felt beyond the boundaries of the technological chain of each process of the blasting cycle considering the volume of broken rock mass involved and of the given physical and geometrical parameters. Therefore, within the framework of the national economy it is proper to consider the final product of blasting as it is directly affected by a blast.

In practice, the marketable mineral, of a particular size and geometrical parameters, measured in tonnes or cubic metres, is termed the final product. If in the process of continued usage of this product, the influence of subsequent processes is felt, then depending on the extent of resources spent it is necessary to take into account the effectiveness of all subsequent processes as well.

While assessing the seismic effects of a blast, the twin requirements of a blast should be considered: On the one hand obtaining a product with given normative specifications, while on the other ensuring seismic-safe conditions

---

[1]This chapter was written in collaboration with L.A. Medvedev.

Therefore, it is necessary to distinguish these two principal factors in the technoeconomic calculations (TEC). The ratios may differ in each specific case and solving the problem of seismic safety can assume the highest priority. In any event, the introduction of measures to reduce seismic effects on the rock massif is determined by the technoeconomic criteria of evaluation (TECrE). The maximal difference between actual and normative profit obtained from the scale of the end-product of blasting operations, serves as the TECrE [22].

In accordance with the guidelines and methodological instructions [23], the maximal value of the TECrE is calculated by the formula

$$E_a = (Pr - E_{norm}K_{sp})Q_{ep} = max, \qquad \ldots (6.1)$$

where E_a is the annual economic effectiveness obtained from the sale of the end-product, roubles/year;

     Pr is the actual specific profit or growth in profit $Pr_1 - Pr_2$ achieved by the enterprise from selling the end-product, rouble/unit of end-product;

     Pr_1 is the profit realised out of the sale of the high-grade product obtained by using measures to counter seismic effects (according to the new variant), roubles;

     Pr_2 is the profit realised out of the sale of poor (earlier) grade (according to the base variant), roubles;

     E_{norm} is the normative coefficient of economic effectiveness of utilising capital investment, taken as equal to 0.15;

     K_{sp} is the actual specific resources (material, labour, financial) spent on the blasting cycle as a whole or on measures to counter seismic effects alone, roubles/unit of end-product;

     Q_{ep} is the volume (quantity) of end-product, units.

The term being subtracted in equation (6.1), namely $E_{norm}K_{sp}$, indicates the normative profit, which should be ensured while undertaking any activity, including those for countering seismic effects.

The maximal value of TECrE or the difference according to formula (6.1) makes it necessary to use the graphoanalytical method, based on which the economically viable or permissible boundaries of outflow of resources, for introducing in a mine measures that reduce seismic effects as well as the expenditure on removing the consequences of blast-induced seismic effects on the rock massif and surrounding objects, are determined.

It is evident that expenses on scientific-research work, experimental-design works and those related to the introduction of the research results in a mine are also to be included in the total expenditure under seismic protection measures.

As the technological process of mining is the basis for economic evaluation of blast-induced seismic effects on rocks, the determination of permissible boundaries of outflow of resources for protective measures forms a part of the entire blasting cycle.

All technological processes in the preparatory blasting cycle (PBC) are mutually related, yet each has a distinct significance in terms of economic evaluation of expenditure and effectiveness. Therefore, it is advisable to know the general index that could determine the effectiveness of the entire process as well as of each unit process (or measures) satisfying the conditions of correlation between the base and new variants [31]:

——According to the volume produced by the new production techniques;

——According to qualitative parameters that reflect physicomechanical properties, geometrical parameters and product grading;

——Time for completing the correlated jobs;

——According to the social factors involved in the production and utilisation of the end-product, considering their effect on the surrounding medium.

The ratio between total amount spent on resources and the scaling coefficient calculated as per the following formula, is taken as the general index of comparative and relative evaluation of the unit process and the entire cycle [23]:

$$\alpha_t = (1 + E)^t, \qquad \qquad \dots (6.2)$$

where E is the scaling norm assumed as equal to 0.1;

t is the number of years, reckoned from the beginning of the accounting year to the given year (distinguishing in particular the expenditure and results of the given year).

In practice, the actual profit of an enterprise is determined as the difference between price of the end-product and its production cost and is calculated as per the formula

$$\{ \qquad Pr_f = Pce - \Delta C, \qquad \qquad \dots (6.3)$$

where Pr_f is the actual profit realised from a unit of end-product, roubles;

Pce is the price of the unit product, roubles;

ΔC is the increment in production cost of unit product, roubles.

The cost involved in blasting operations depends on many factors, including that of protective measures. As already mentioned, blast-induced seismic effects are significant in two ways: firstly, they help in producing the end-product (positive effect) and secondly, they exert harmful effects on the near-by surface and underground structures (negative impact). Consequently, the costs of protection measures and costs of enhanced production of mineral (all other indexes remaining constant), should finally cover all costs in eliminating the damaging effects induced by blasting.

By perfecting the location of charges in a massif and hook-up layouts, we can achieve maximal effectiveness of protection measures with minimal expenditure of resources. This can be achieved by analysing the patterns of blast-induced seismic effects on the rock massif. The complex effect of blasting parameters on rock is economically assessed by the total cost of principal technological processes. In doing so, the variations in adjacent sections and technological processes are also taken into account, as they are interrelated technologically.

In practice, the introduction of a new technology involves various measures: rational charge layouts in massif; optimal intervals of SDB; consideration of the anisotropy of rocks; perfection of the charge design; usage of new types of explosives, drilling machines, other instruments and devices etc.

These measures induce a change in the following cost elements for: drilling holes ΔC_d; charging and stemming of holes ΔC_{ch}; sorting out boulders ΔC_{bo}; completion of toe of the bench ΔC_t; explosives and other materials consumed ΔC_m.

Any change in the technology of drilling and blasting operations (DBO) affects the granulometric composition of the broken rock mass and parameters of the broken rock heap. This in turn affects the operational efficiency of loading-transporting media and crushing-sorting equipment. In consequence, the indexes of the next processes should increase and their output should be enhanced: in loading ΔC_{lo}; in transporting ΔC_{tr}; in mechanical crushing ΔC_{cr}.

Here, by end-product is meant the yield of standard-size rock fragments and level of working toe of the bench, i.e., the necessary conditions for normal conduction of the next cycle of DBO.

Thus the total variation in cost of the end-product in the DBO cycle, taking into account adjacent technological operations in the process of preparing saleable mineral, is computed by

$$\Delta C = \Delta C_d + \Delta C_{ch} + \Delta C_{bo} + \Delta C_m + \Delta C_{lo} + \Delta C_{tr} + \Delta C_{cr}.$$

The consideration of increment in cost ΔC in formula (6.3) predetermines a systematic graphoanalytic method of finding the optimal value of actual profit Pr_f and enables one to establish the rational parameters of DBO. For this purpose, it is appropriate to analyse the structure of cost increments in the complex of DBO.

6.2 Change in Structure of Principal Cost Elements in DBO

The input indices for calculating the drilling cost are: specific volume of drilling needed for breaking 1 m^3 of rock mass with specific parameters in the base variant Y_{1d}, m/m^3; same, as per the newly introduced variant Y_{2d}, m/m^3 drilling cost per metre Pce_d, roubles/m.

In case of reduced volumes of drilling, without compromising on the results of blasting, the variation in the cost of unit product can be determined by

$$\Delta C_d = (Y_{1d} - Y_{2d})Pce_d.$$

If the volume of drilling remains constant, but the cost of drilling C_d (roubles/m^3) changes, then the specific increment of this cost is computed as the difference between specific drilling cost in the base variant (Pce_{1d}) and in the new variant (Pce_{2d}), i.e.,

$$\Delta C_d = Pce_{1d} - Pce_{2d}.$$

In case of variation in the costs of charging and stemming in the base and new variants, their specific value as per the end result (roubles/m^3) is computed in a manner similar to the drilling process,

$$\Delta C_{ch} = Pce_{1ch} - Pce_{2ch},$$

where Pce_{1h} and Pce_{2h} are cost of charging and stemming, in the base and new variants, respectively, roubles/m^3.

The input parameters of variation in the yield of boulders are: Y_{1bo} — average yield index of large rock fragments as percentage of total broken volume in the base variant; Y_{2bo} — same, in the new variant; Pce_{bo} — cost of sorting 1 m^3 of product.

Specific variation in the cost of unit product (roubles/m^3) as a result of change in the volume of boulders, is calculated using the equation

$$\Delta C_{bo} = \left(\frac{Y_{1bo} - Y_{2bo}}{100}\right) Pce_{bo}.$$

The change in volume of work pertaining to the working of the bench toe requires the following input data:

Y_{1t} — average per cent of overworking of bench toe during blasting in the base variant;

Y_{2t} — same, in the new variant, due to the usage of new or better explosives, blasting layouts, charge design;

Pce_t — cost of separating 1 m^3 of toe in specific field conditions.

The specific variation in the cost per unit of product (roubles/m^3) as a result of better overworking of toe is given by

$$\Delta C_t = \left(\frac{Y_{1t} - Y_{2t}}{100}\right) Pce_t.$$

If usage of new types of explosives results in reduced specific consumption of explosives, their cost remaining constant, then the specific change in production cost can be calculated as follows:

$$\Delta C_m = (Y_{1m} - Y_{2m})Pce_m,$$

where Y_{1m}, Y_{2m} are specific consumption of explosives, kg/m^3 of blasted rock, produced in the base and new variants respectively; Pce_m is the purchase price of explosives, roubles/kg.

If the specific consumption of explosives as well as their price vary, then the specific change in the cost of unit product (roubles/m^3) is determined from the equation

$$\Delta C_m = Y_{1m}Pce_{1m} - Y_{2m}Pce_{2m},$$

where Pce_{1m} and Pce_{2m} are material cost in the base and new variants respectively, roubles/kg.

When new or better types of explosives are used with various additives, the price of explosives (roubles/kg) is given by

$$Pce_m = Pce_{1m}n_1/100 + Pce_{1a}n_2/100 + Pce_{ad},$$

where n_1, n_2 are percentage of base explosive and additives respectively in the recommended composition; Pce_{1a} is the price of additive, roubles/kg; Pce_{ad} are additional expenses incurred for manufacturing the new explosive by means of additives or mixing.

Any change in the indicated technological parameters influences the adjacent processes of loading and transporting operations.

The technoeconomic effectiveness of loading operations can be enhanced by reducing the yield of boulders in the broken rock mass. This is evaluated by a corresponding variation in shovel output. The input quantities for computation are: P_1 and P_2—shovel output in the base and new variants respectively; Pce_{1ms} and Pce_{2ms}—cost incurred in one shovel shift in the existing operating conditions, in the base and new variants respectively, roubles.

The specific change in loading cost per unit product (rouble/m³) is calculated by the formula

$$\Delta C_{lo} = Pce_{1ms}/P_1 - Pce_{2ms}/P_2.$$

Shovel output per shift (m³/shift) is given by

$$P_s = 3600E_bT_sK_{fill}R_{ut}/(K_{loose}t_{cycle}),$$

where E_b is the bucket capacity of shovel, m³;

T_s is the duration of shift, hr;

$K_{fill} \approx 0.9$ is the utilisation coefficient of bucket volume-wise (rock-filling);

$R_{ut} \approx 0.7$ is the utilisation coefficient of shovel during a shift;

K_{loose} is the coefficient of rock loosening;

t_{cycle} is the duration of one shovel cycle, min.

The values of K_{loose} and t_{cycle} could be calculated according to the empirical relationships:

$$K_{loose} = 1.1 + 0.36Q_{bo};$$

$$t_{cycle} = 43.6 + 1.2Q_{bo},$$

where Q_{bo} is the average percentage of boulders in the total volume of broken rock.

Output of shovel can be determined, taking into account the coefficient of increase in output (considering reduced yield of boulders), by

$$K_{op} = \frac{1 + 0.045Q_{1bo}}{1 + 0.045Q_{2bo}},$$

where Q_{1bo} and Q_{2bo} are average volume-wise yield of boulders in the base and new variants respectively.

In practice, the shovel output is also determined by considering the granulometric composition of the rock mass (uniformity of fragmentation) and heap of broken rock parameters.

As an example, the loading cost of 1 m³ broken rock by EKG-4.6 shovel can be computed by the empirical relationship

$$\text{Pce}_{1o} = \frac{C_{ms}(1.1 + 2d_{av})(60 + 0.55H_h^2 + 67.2d_{av})}{210E_bT_sR_{ut}\sqrt{15H_h}},$$

where C_{ms} is the cost of a single shovel shift, roubles;

d_{av} is the average diameter of rock fragment, m;

H_h is the average height of heap of broken rock, m.

The output of in-pit transport also varies if there is a precondition for qualitative fragmentation of rocks, taking into account the blast-induced seismic effects on the massif. This is reflected in the complete and even loading of transport media and rate of their turnaround. It is advisable to adopt direct computation for determining the variation in transport costs.

The effectiveness of crushing-sorting equipment also depends directly on the granulometric composition of the broken rock. In this case it is determined by primary crushing and is assessed by the average diameter of rock fragment. The input data for calculation: Pce'_{1cr} and Pce'_{2cr} — costs of primary crushing of the rock mass, in the base and new variants respectively roubles/m³.

The specific change in cost of unit product (roubles/m³) (in terms of crushing) is given by

$$\Delta C_{cr} = \text{Pce}'_{1cr} - \text{Pce}'_{2cr}.$$

The crushing cost (roubles/m³) is calculated empirically by

$$\text{Pce}'_{cr} = \frac{\text{Pce}'_{cr.ms}}{(160 - 114d_{av})T_sR_{cr.ut}},$$

where $\text{Pce}'_{cr.ms}$ is the cost of single crusher-shift, roubles;

$R_{cr.ut}$ is the utilisation coefficient of crusher (taken as 0.7 on average).

Besides the above factors that induce variation in technoeconomic indices (TEI), seismic protective measures in quarries have special significance. Their distinctive feature is that expenditure in realising them can be reduced even more than for the base variant.

Seismic-safe blasting enables enlargement of the scale of blasting, and a better quality of rock fragmentation, which in turn results in reduction of downtime of the entire quarry equipment, reduction in duration of preparations and repairs within the quarry and in surrounding objects associated with the blast, reduction in costs of creating protective structure and repairs of surface and underground structures disturbed by vibrations.

Reduced duration of downtime due to fewer blasts results in economic returns (roubles/year):

$$E_a = C_{raw}\frac{C_{fc}}{100}Pr_h(t_1 - t_2),$$

where C_{raw} is the cost of mining 1 T of raw mineral, roubles;

C_{fc} are conditional fixed costs, measured in percentage of unit product cost (for quarries C_{fc} is taken as equal to 40–50%);

Pr_h is the hourly output of the quarry, T;

t_1 and t_2 are scheduled downtimes in the base and new variants, hr.

From the above analysis of increment in total cost, it follows that the optimal cost indices of the entire drilling-blasting cycle as well its unit processes should be determined taking into account the incremental cost (profits) or finally the maximal effectiveness of expenditure computed by means of (6.2).

With this aim in view, a graph interpretation of (6.3) is shown in Fig. 63,a, for the condition of uniform increase in the volume of production, Q_{ep}, with specific quality in the natural or cost expression. The ultimate optimal value of volume of productivity of mining and benefication plants, as is well known, is determined by several factors, major among which are confirmed reserves, life of mine and mineral grade — all of which are of decisive significance for the national economy.

As can be seen from Fig. 63,a, with an increase in volume of production, the price, Pce, of realised product increases proportionally (curve 1), which

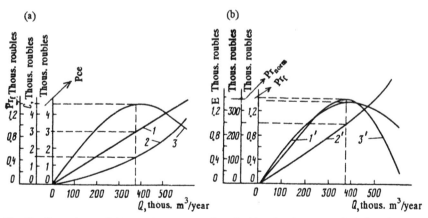

Fig. 63: Dependence of (a) actual expenses and profit (b) and technoeconomic indices based on final product on output of the mine:

1 — price of realised product; 2 — cost of mining; 3 and 1' — actual profit; 2' — data about normative profit, characterised by the product $E_{norm} K_{sp}$; 3' — technoeconomic effectiveness.

usually does not vary within a year. Consequently, with a uniform rise in volume of production, the price of the entire realised product also rises uniformly. In specific cases this growth in price can be expressed by a direct equation

$$\Delta \text{Pce}_{\text{index}} = a_{\text{pce}} \cdot Q_{\text{ep}}, \qquad \dots (6.4)$$

where $\Delta \text{pce}_{\text{index}}$ is the index showing the price rise of the entire realised product, roubles;

a_{pce} is the empirical coefficient of price rise, depending in each concrete case on the complex conditions of production (dimensionless).

The cost index C of the product (curve 2) has a tendency to rise with an increase in volume of production. This is explained by rise in specific costs associated with complications of geo-mining conditions, parameters of mining methods and reduced conformity of properties of product to specifications. Consequently, costs will constantly rise if there is even (uniform) growth in volume of production.

Curve 3 in Fig. 63,a has a hyperbolic shape and can be expressed by

$$\Delta C = \frac{Q_{\text{ep}}}{b + cQ_{\text{ep}}} \text{ roubles}. \qquad \dots (6.5)$$

where b and c are empirical coefficients that characterise technoeconomic conditions of production.

The actual profit, Pr_{f}, from realisation of the product, according to formula (6.3), is equal to the difference in price rise, $\Delta \text{Pce}_{\text{index}}$ (obtained from equation (6.4)) and cost, ΔC (obtained from equation (6.5)). This difference is represented by curve 3 in Fig. 63,a. As is evident, curve 3 exhibits a tendency towards reduction with an increase in volume of production. This conforms to reality since, when a mine achieves ultimate output, the actual profit index stabilises and subsequently may slide into a negative value.

The economic effectiveness, according to formula (6.1), is not only determined by the level of actual profit, but also by the subsequent stages of mineral utilisation.

In accordance with the methodology [23], it is necessary to consider also the normative profit $E_{\text{norm}} K_{\text{sp}}$, which characterises the approach (with corresponding scale of national economy) to the evaluation of economic effectiveness of the measures introduced (techniques, technologies and recommendations to counter seismic effects etc.). It is obvious that with a reduction in expenditure on resources for undertaking the new measure (i.e., to counter seismic effects), the national economy will prosper. However, with an increase in volume of production, firstly the marketable resources would grow and the difference within brackets in formula (6.1) would decrease.

The variation in actual profit Pr_f (curve $1'$) and normative profit $E_{\text{norm}}K_{\text{sp}}$ (curve $2'$) is shown in Fig. 63,b. Their difference is represented by curve $3'$. The maximal value of profit in this case is obtained at a production of 375,000 m^3 per annum and conforms to the optimal values of principal TEI (price of re-alised product amounts to 3000 roubles, increment in cost 1500 roubles and the structural elements of its components[*]). These optimal values of productive ex-penditure of resources just happen to be the maximum permissible and are also technoeconomically justifiable.

The economic effectiveness calculated by formula (6.1) incorporates an evaluation of all the production processes involved in obtaining the final product. Practicality demands anticipation of the economic effectiveness of any measure or any expenditure likely to be incurred in obtaining and realising the final product. Hence to achieve overall economic effectiveness, the fractional partici-pation of each of the measures associated with the perfection of techniques and production technology ought to be determined. Little heed has been given to this problem in the published literature. In the instructions published for estab-lishing the economic effectiveness of capital investments in civil engineering, it is mentioned that if duration of completing individual tasks or stages of work are reduced, resulting in giving the premises for occupation within a short time, then the effect associated with completing these types or stages of tasks is de-termined from the proportion of costs of the particular tasks or stages in the total cost of the building.

In implementing the foregoing, it should be noted in addition to capital expenditure, the costs incurred in research and development work (including trial of equipment and implementing new recommendations) should also be considered.

It follows from the above that the fractional economic effectiveness as per the final result of implemented measure should be determined, taking into account the proportionality coefficient of the effect of fractional cost of each measure, given by the formula

$$E_{\text{a.m}} = (\text{Pr}_f - E_{\text{norm}}K_{\text{sp}})A_2 K_{\text{prop}}, \qquad \ldots (6.6)$$

where $E_{\text{a.m}}$ is the economic effectiveness of the given measure per annum, roubles;

K_{prop} is the proportionality coefficient of economic effect due to the im-plementation of the given measure of perfection or development of a new variant.

The foregoing principle of cost-based evaluation does not reflect completely the complexity of management effectiveness, since there may be cases of non-conformity to this principle. Therefore, it is imperative to consider the time

---

[*] sic; no figure given in Russian original — General Editor.

factor in the coefficient K_{prop}, which characterises the intensity of spending of resources, i.e., expenditure related to a specific time. The index K_{prop} is computed by the formula

$$K_{prop} = \frac{Z_{base}}{Z_{new}}, \qquad \ldots (6.7)$$

where Z_{base} and Z_{new} are total resources spent for implementing the measure in the base and new variants respectively.

Measures aimed at reducing blast-induced seismic effects enable augmentation of ore reserves for exploitation in mines. The quantum of this increase could be quite considerable and of notable technoeconomic significance by permitting not only exploitation of additional mineral resources, but also utilisation of the funds in a productive manner, which ultimately would extend the life of the mine. The quantum of additional resources (brought into exploitation) in each case, is established taking into account specific complex conditions of mining.

A decisive role is played by direct expenditure (cost) and average realised cost (price) of product conforming to the consumer's needs, in evaluating the effectiveness of mining (taking into account the added reserves and protective measures undertaken).

The economic effectiveness from the point of view of the national economy is computed by the formula

$$E = (Pce - C)A, \qquad \ldots (6.8)$$

where Pce is the average cost (price) of realised product, roubles;

C is the average cost of producing product after reserves are added, roubles;

A is the volume of additional reserves of raw minerals, m^3.

Whenever protective measures require additional funds, the economic effectiveness is computed by the formula

$$E = (Pce - C)A - Z_{ad}, \qquad \ldots (6.9)$$

where Z_{ad} is the additional cost for taking protective measures, roubles.

The considerable experience gained in rock breakage by blasting calls for a comprehensive evaluation of protective measures taken to counter seismic effects since blasts have become larger in scale. In constructing different objects by means of blasting, the major factor of effectiveness is the significant reduction in duration of operations (by consuming a large quantity of explosives), which contradicts the conditions for seismic safety. The economic effectiveness in such conditions is computed by the formula [14]

$$E = (E_{con} - E_{expl})Z_{ad}, \qquad \ldots (6.10)$$

where E_{con} is the effectiveness at the construction stage of ground structures, roubles;

E_{expl} is the effectiveness at the exploitation stage, roubles.

At the construction stage of the object, savings are made from the conditionally fixed part of overhead expenses, which is computed by the formula

$$E_y = H_{oh}(1 - T_2/T_1), \qquad \ldots (6.11)$$

where H_{oh} is the conditionally fixed part of overhead expenses, roubles;

T_1 and T_2 are duration of completing the process in the base and new variants respectively, days.

At the exploitation stage, due to reduced duration of completing the operation, a one-time saving is achieved:

$$E_{expl} = EF_{pr}(T_2 - T_1), \qquad \ldots (6.12)$$

where F_{pr} are productive funds commissioned before scheduled time, roubles.

The resources spent on undertaking protective measures are the sum of the cost of constructing additional structures: equipment, instruments and devices used etc. However, in some cases the outlay on protective measures could completely wipe out the effectiveness achieved through other measures. Therefore the technoeconomic justification for undertaking protective measures should precede the project preparation.

6.3 Results Obtained After Introduction of Protection Measures against Seismic Effects

Let us take the example of effectiveness achieved by introducing measures for seismic safety in the Skala-Podol'sk limestone quarry of GPO 'Dorstroimateriali' (Materials for Road Construction) under the Ministry of Road Construction, the Ukraine. This quarry works in very congested geo-mining conditions, with (rural) settlements in the north and south in close proximity not to mention the industrial buildings of the asphalt-concrete factory. Development of a mining front in the direction of the aforesaid structures was temporarily stopped by the State Mining Inspectorate until special investigations were conducted and recommendations given for conducting safe blasting operations in the quarry. The Geophysics Institute of Academy of Sciences, Ukraine, conducted experimental investigations in the quarry and offered recommendations [4–7] which made possible conduction of blasting operations annually and enabled exploitation of mineral resources in zones at distances of 200 m from residential houses and 70 m from industrial buildings without incurring any capital expenditure for repair and maintenance of these structures.

Based on experimental investigations and contractual conditions, the quarry management was provided with 'Recommendations for safe blasting operations at the western flank of Skala-Podol'sk limestone quarry' for mining in the above-mentioned zones. The annual TE effectiveness was computed in conformity with these recommendations.

White gravel (40–70 mm), asphalt-concrete, black gravel and mineral fines are produced in the Skala-Podol'sk Asphalt-Concrete Factory.

According to the existing (base) technology of DBO, the seismic effects on the rock massif as well as on surrounding structures, in particular on the factory buildings, exceeded the limits of permissible norms. Thus the need arose for stopping mining operations and for conserving part of the exploitable reserves.

According to the recommendations on advanced by the Geophysics Institute, protective measures were taken. These included a particular sequential working of sections, parameters of drilling and blasting charges, differentiated selection of delay interval between charges, creation of screening cavities in the massif whose location was determined by the mutual relationship between parameters of the block and properties of the massif etc.

The principal input data for computing just the economic effectiveness of the recommendations suggested for mitigating seismic effects, are given in Table 17.

In the base variant, when the factory was required to close down, only white gravel of 40–70 mm fractions and mineral fines were being produced, while in the new variant asphalt-concrete and black gravel were also produced.

Since the expenditure on producing all four types of products was lower than their cost, the factory improved its profitability. The economic effectiveness of seismic measures was computed according to formula (6.6). In doing so, the actual profit gained by the factory (roubles/year) was determined by comparing the base and new variants by means of the formula

$$\mathrm{Pr_f} = \mathrm{Pr_2} - \mathrm{Pr_1},$$

where $\mathrm{Pr_1}$ and $\mathrm{Pr_2}$ are profits gained by the factory, in the base and new variants respectively, roubles/year.

Profit in the base variant is given by

$$\mathrm{Pr_1} = [(\mathrm{Pce_g} - C_g)A_g] + [(\mathrm{Pce_{mf}} - C_{mf})A_{mf}],$$

where $\mathrm{Pce_g}$ and $\mathrm{Pce_{mf}}$ are price of 1 T of gravel and mineral fines respectively, roubles;

Table 17

Product	Annual production, '000 T	Cost, roubles	Price of unit product, roubles
White gravel of 40–70 mm fractions	60*	10,671	2.4
Asphalt-concrete	155	882,460	7.2
Black gravel	195	1,000,025	6.55
Mineral fines	95	533,790	8.0

* In 000's of cubic metres.

C_g and C_{mf} are production cost of 1 T gravel and mineral fines respectively, roubles;

A_g and A_{mf} are volume of production of gravel and mineral fines, respectively, T.

The price of each product is given by

$$Pce_g = Pce_{g.u}\, A_g, \text{ roubles;}$$
$$Pce_{mf} = Pce_{mf.u}\, A_{mf}, \text{ roubles}$$

where $Pce_{g.u}$, $Pce_{mf.u}$ are unit price of gravel and mineral fines respectively, roubles/T.

Substituting the input data in the formula for computing Pr_1, we get:

$$Pr_1 = (60,000 \times 2.4 - 10,671) + (95,000 \times 8 - 533,790) =$$
$$= 133,329 + 2,266,210 = 359,539 \text{ roubles/year } [sic\,!].$$

Thus, the annual profit of the factory in the base variant amounts to 359,539 roubles.

In the new variant, the profit gained from realising the products is given by:

$$Pr_2 = [(Pce_g - C_g)A_g] + [(Pce_{as} - C_{as})A_{as}] + [(Pce_{bg} - C_{bg})A_{bg}] +$$
$$+ [(Pce_{mf} - C_{mf})A_{mf}] = (60,000 \times 2.4 - 10,671)$$
$$+ (155,000 \times 7.2 - 882,460) + (195,000 \times 6.55 - 1,000,025)$$
$$+ (950,000 \times 8 - 533,790) = 13,332 + 233,540$$
$$+ 277,225 + 226,210 = 870,304 \text{ roubles/year } [sic]$$

The relative profit evaluated as per the new variant is $Pr_f = Pr_2 - Pr_1 = 870,304 - 359,539 = 510,765$ roubles/year.

As already mentioned, the effectiveness attributed to DBO is determined taking into account the proportionality coefficient computed by formula (6.7).

The cost of DBO and research and development work in the Skala-Podol'sk quarry amounted to 112,700 roubles. This cost was calculated as a fractional ratio to the total cost of operations, which was about 252,770 roubles.

Consequently

$$K_{prop} = \frac{112,700}{252,770} = 0.45.$$

In the new variant of mining operations, there is no need for additional capital investment. Therefore the subtracted term $E_{norm} K_{sp}$ in formula (6.6) is deleted.

Thus, the economic effectiveness of introducing the recommendations on protective measures (which are realised in the DBO), is:

$$E_{a.m} = 510,765 \times 0.45 = 229,800 \text{ roubles } [sic].$$

Therefore, the annual cost of DBO amounting to 112,700 roubles is recovered within

$$T_0 = \frac{112,700}{229,800} \approx 0.5 \text{ year.}$$

The optimal values of productive costs and the economic effectiveness achieved in such a case are established for the final product, in the same manner as above.

The recommended graphoanalytical method of finding the optimal (economically justified and permissible) outlay of resources and their components, takes into account the feedback from the input (design) parameters for fixing the maximal annual economic effectiveness and conversely, for correcting the input (design) parameters and their optimal values.

The components of outlay of resources are contained in the major process of DBO and are computed from the final product on the basis of intermediate measurements. Thus in preparing the area for drilling, the volume of work involved is measured in square or cubic metres and is grouped with the final product (m^2/m^3 and m^3/m^3) in drilling of holes in metres (m/m^3), in charging in kilograms (kg/m^3) etc.

In field conditions all processes in the cycle are interrelated and yet the need arises to assess objectively, based on the final product, each unit process or operation. Hence it is essential to establish a generalised (besides specific) index, capable of concomitant TE evaluation of local and global aspects. The intensity of outlay of resources (rate of expenditure) Z_{in}, roubles/unit time, can be taken as such an index. The average index of the entire drilling-blasting cycle is correlated with Z_{in}.

The calculated data of intensity of outlay of resources, expressed in units of cost and time are given in Table 18.

Table 18

Major processes	Duration of processes		Cost		Rate of expenditure	
	Min	%	Roubles	%	Roubles/min	%
Site preparation for drilling	150	2	5	0.14	0.0009	0.01
Haulage of drill rig to drilling site	20	0.3	0.6	0.017	0.030	0.49
Drilling	5250	85.2	420	11.9	0.080	1.0
Separating boulders	322	5.2	25	0.714	0.078	1.0
Charging and blasting of holes and inspection of blast site	420	7.3	3050	87.229	7.262	97.5
Total for entire cycle	6162	100	3500.6	100	7.4509	100

It follows from the table that the drilling process is the most time-consuming at 85.2% of the duration of the entire cycle extending up to 6162 min (taken as 100%).

It is essential to know the cost of each of these processes for the purpose of evaluating their effectiveness. The data obtained indicate that the most costly process is that of charging, blasting and post-blast inspection of site. This amounts to 87.2% of the total cost of the entire cycle. In this the expenditure on materials is included, amounting to 98.4% in the given case of the total resources spent in the entire process.

From this brief analysis of the rate of expenditure for an entire drilling-blasting cycle, it becomes evident that the blasting process was the most effective, as past experience had largely been used. It is advisable therefore to achieve further perfection of the blasting cycle and to enhance its effectiveness by improving the mechanism of blast effects on the massif while implementing protection measures as outlined above.

The recommendations developed here have been implemented in various mines and significant economic effectiveness obtained as a result of introduction of the protection measures (Table 19).

The major principles for evaluating the protection measures can be formulated in the following manner:

Due to the close interrelation and influence of all processes and operations of the blasting cycle, the TE evaluation of each of the component processes should be done based on the final result — economic effectiveness of realising the final product. Any measures, including the protective ones, are better analysed through feedback; from the design (input) indices to the maximum economic effect and optimal values of input data contained in it, which are later corrected in the process of optimisation.

Local assessment of processes (measures) can be done based on the rate of expenditure. The evaluative criterion in this case is the index of rate (intensity) of expenditure in the entire blasting cycle. It is obvious that the effectiveness of individual processes and consequently their subsequent perfection is directly proportional to the value of this index.

The method of determining permissible (economically justifiable) expenditure of resources is an appropriate one, which considers only the evaluation principle. The existing concepts for enhancing TE effectiveness through recommendations for protection measures are valid in this method.

6.4 Physical Effectiveness of Industrial Blasts
(written in collaboration with A.I. Kondrat'ev)

In any technological process the nature and content of the final objective vary. The effectiveness of a specific process can be correctly assessed only on the basis of its contribution towards achieving the overall objectives of the entire process.

Table 19

Quarry, Department	Measures introduced, year of introduction	Purpose of blasting	Object under protection	Economic effect due to introduction of measures '000 roubles
(1)	(2)	(3)	(4)	(5)
Popel'nyanskii quarry, GPO 'Dorstroimaterialy' (Materiaals for road construction). Ministry of Road Construction, Ukraine.	Screening (offering seismic protection) partitions, 1970	Preparation of rock mass	Plant area and buildings	100*
Gaivoronskii, Rakitnyanskii, Popel'nyanskii quarries, GPO 'Dorstroimaterially', Ministry of Road Construction, Ukraine.	Same as above, 1971–1982	Same as above	Same as above	36.1
Chikalovskii quarry, GPO 'Dorstroimaterialy'; Ministry of Road Construction, Ukraine	Same as above, 1972–1973	Same as above	Same as above	1500*
Samenskii quarries KDZ of 'Granitas' group	Same as above, 1974	Same as above	Same as above	30*
Polonskii, Rokitnovskii, Starokonstantinovskii, Skala-Podol'sk quarries, GPO 'Dorstroimaterialy', Ministry of Road Construction,Ukraine	Seismic safe parameters taking into account anisotrophy of rocks, 1978–1980	Same as above	Residential buildings and industrial area of the factories	167.8

159

Organization	Technology	Application	Location	Value
'Karpenyai' quarry of 'Granitas' group	Seismic safe construction of charges on igdanite base, 1979–1982	Preparation of rock mass	Industrial area of Cement Factory	63*
Kamenets-Podol'sk quarry of ZAB GPO 'Dorstroi-materialy' Ministry of Road Construction, Ukraine	Seismic safe parameters of large-scale blasts taking into account overburden rocks, 1980	Same as above	Deep gorge of Smotrich River	34*
Quarry No. 2 of north-eastern section of Vodinsk deposit, Kuibyshev Sulphur factory, 'Soyuz Sera' group, Ministry of Fertilisers, USSR	Seismic safe blasting technology, 1981–82	Same as above	Trunk gas line	65.25
'Yuzhgiprostroi', Ministry of Industrial Construction Materials, USSR	Seismic safe parameters of large-Scale blasts taking into account over-burden rocks, 1982	Same as above	Deep gorge of Smotrich River	Utilised in reconstruction project of Kamenets-Podol'sk Asphalt-Concrete Factory
Skala-Podol'sk quarry ZAB GPO 'Dorstroimaterialy', Ministry of Road Construction, Ukraine	Seismic safe blasting technology (200 m zone), 1982	Same as above	Buildings of northern, southern settlements and industrial area of factory	1985.25

(Contd.)

Table 19. Continued

(1)	(2)	(3)	(4)	(5)
'Shatlykgazstroi' Trust Ministry of Oil and Gas Industry, USSR	Seismic safe parameters of SDB and screening, 1982	Compaction of loess type subsided soils	Kazanzhiksk station	238
Polonnoe quarry, Bryansk 'Promstroimaterialy' group, Ministry of Industrial Construction Materials, USSR	Seismic safe weights of explosive charges, 1983	Preparation of rock mass	Trunk gas pipe line	160
Skala-Podol'sk quarry ZAB GPO 'Dorstroimaterialy', Ministry of Road Construction, Ukraine	Seismic safe blasting technology 1983	Same as above	Buildings of northern, southern settlements and industrial area of factory	1877
Podgornoe quarry of Zakarpat, Ministry of Road Construction, Ukraine	Seismic safe parameters of SDB, 1983	Same as above	Trunk gas line	104
Krutikhinsk PMK of 'Kamen'vodstroi' Trust, Irrigation Ministry RAFSR	Seismic safe parameters for SDB screening, 1984	Compaction of subsided soils below foundations of buildings	Grain drying plant, residential houses, ground communication cable	132.7
Central quarry of Yazovskii mine, Yavorovskii RPO 'Sera', Ministry of Fertilisers, USSR	Seismic safe blasting operations, 1984	Preparation of rock mass	Sanatorium 'Shaklo' and administrative building for enterprise	Incorporated into project
Podorozhnskii mine of Rozdol'skii PO 'sera' group, Ministry of Fertilisers, USSR	Seismic safe charge weight per delay and for entire blast, 1984	Same as above	Mine surface plant and secondary school	7*

				Incorporated into project
L'vov branch of 'VNIPIsera', Ministry of Fertilisers, USSR	Predictive evaluation of seismic safe blast parameters,1984	Preparation of rock mass	Pumping station of central quarry of Yazovsk mine	
Skala-Podol'sk quarry of ZAB GPO 'Dorstroi-materialy', Ministry of Road Construction, Ukraine	Seismic safe parameters of SDB, screening (70 m zone), 1984	Same as above	Factory buildings	229.8
Podorozhnenskii mine of, Rozdol'sk PO 'Sera', Ministry of Fertilisers, USSR	Seismic safe charge weight per delay 1986	Same as above	Mine surface plant and secondary school	47
Berestovetsk quarry of 'Rovnodorstroimaterialy' Ministry of Road Construction, Ukraine	Same as above, 1987	Same as above	Rural cemetery, secondary school	168
Northern quarry of Rozdol'sk PO 'Sera', Ministry of Fertilisers, USSR	Same as above,	Same as above	Residential houses of Malekhovo village	59

*Expected economic effect.

During any technological process the labour requirements undergo change under a specific set of conditions. To reflect the link between this set of conditions and the results of the technological process the concept of effectiveness is used. The task of researchers lies in determining the optimal relationship between the conditions created and the production results. An economic analysis of the technological process expressed in equivalent cost categories helps in measuring effectivity directly, as the ratio of pure result (after deduction of costs) to the total expenditure (expressed in cost form).

All economic categories that characterise the production processes are founded on real physical processes. But the concept of effectiveness does not exist in the physical approach. To study human activity intended for achieving specific targets, this concept is essential. The entire organisational system for production that ensures effectivity is a reflection of the knowledge and experience gained by society at a given moment.

Each production process possesses concomitantly a physical and informative nature. Information plays a decisive role in forming relationship between expenditure of various types and the result obtained, in other words on the degree of process effectiveness. Information flow depends on the organisation and structure of the production process.

In evaluating the quality of a product, reference is made to the complex ratios of properties or characteristics of technical systems; economic indices and assessment of effectiveness happen to be only approximate measures for expressing them.

The so-called functional criteria and their evaluation reflect the quality of fulfilling the task by the production system functions. These criteria do not possess the generality or universality of economic evaluation. But the tasks of functional evaluation are entirely different. They help the engineer and technologist to choose the best variant for implementing a project. The cost and functional criteria should be complementary. The technological criterion, as distinct from its generalised economic counterpart, points to the organisational and technological links among the processes.

The physical processes measurable in physical quantities are the basis for realising the functions of production processes. It is logical, therefore, to assume that all the multifaceted functional evaluations can be systematised during their analysis by physical concepts, allowing their conversion into physical quantities. Then the system behaviour can be determined by interrelations of physical laws and properties of the specific system. Physical assessments and effectivity indices have become the means of revealing the interrelationships and realisation of these linkages in system management.

As applied to production blasts, physical effectivity can be evaluated by the results of granulometric composition of the broken rock governed by the nature of distribution of individual grains in a massif of complex structure. In this case fragmentation results characterise some fraction of expenditure of the total blast energy and are interrelated with all other manifestations (seismic effects etc.). It becomes evident that the greater the degree of rock fragmentation, the lesser the energy spent on seismic effects. The powder factor can be predicted based on the latter aspect.

Evaluation of the quality of rock fragmentation varies over a wide range for the same set of input data. This is explained by errors in determining the parametric values of the massif, their complexity and incomplete consideration of the influencing factors.

While determining the specific consumption of explosives, the average diameter of rock fragments is taken as a reference. Different conditions and time of crack formation cause varying degrees of jointing in the massif. Among the calculated parameters, the values and nature of dispersion of particle dimensions in the massif are absent. The particle dimensions are random values and statistical evaluation is applicable to them. Similar assessments are obtained in the determination of jointing in massifs [25]. To evaluate the degree of homogeneity of a massif, it is adequate to know the share of the prevalent class of particles or the maximum value of differential function of the particle size distribution (Fig. 64,a).

Based on the analysis of large-scale blasts in granite quarries it has been established that the nature of particle size distribution in the massif significantly influences the quality indices, i.e., the yield of non-standard classes. This can be explained by the non-linear relationship between the degree of fragmentation of the massif with the specific explosive consumption in various ranges of particle sizes (in their natural state). The weight of explosive calculated for fragmenting the middle size class helps in the overbreaking of smaller fragments while the same charge weight breaks coarser fragments inadequately. For various particle size distributions, the fraction of the massif subjected to the effect of blasting a charge is also different, whose value does not conform to the calculated one. The more homogeneous the massif, the more accurately the calculated charge gives the required degree of breakage and qualitative indexes. Here the possibility exists for evaluating the choice of series of charges in a large-scale blast through its physical effectivity. Under similar conditions, the best method of rock breakage by blasting provides minimum yield of non-standard fragments and maximum accuracy in the degree of breakage (Fig. 64,b, wherein for clarity distribution functions and not histograms are given). Degree of breakage (crushing)

$$K_{1br} = \frac{d_0}{d_1}; \quad K_{2br} = \frac{d_0}{d_2}.$$

Since one and the same volume of rock mass is being evaluated, let us determine the selectivity of any blasting method

$$K_{1sel} = \frac{p_1(d)_{max}}{p_1(d_0)_{max}}; \quad K_{2sel} = \frac{p_2(d)_{max}}{p_2(d_0)_{max}}.$$

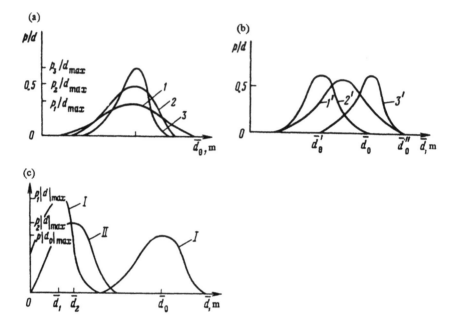

Fig. 64: Distribution of particles in massifs of (a) varied complexity, (b) zone-wise and (c) in broken rock fragments:

1, 2, 3,—degree of complexity of massifs having similar average size of particles, d_0; $1'$—rock massif; $2'$, $3'$—zones of rock massif having values d_0' and d_0''; I, II—average size of rock fragments d_1 and d_2 respectively.

It is obvious that the first method is more effective, as $K_{1sel} > K_{2sel}$.

It should be borne in mind that the task of rock breakage becomes more complex with the reduction in d of standard classes. In such a case the yield of non-standard classes becomes more. A method having the maximum coefficient K_{sel} is required in order to provide a blast with enlarged requirements. The approximate values of K_{sel} for assessing a blasting method can be unity. The distribution of fragment size in broken rocks conforms to the structure of the massif. With $K_{sel} < 1$, scattering of fragments increases while $K_{sel} > 1$ indicates better selectivity of the breakage method.

It is natural to assume that the value of $p(d_0)$ approaching unity (more homogeneous massif as per the share of particles) is readily manageable: the computed parameters of blasting operations should conform to blockiness whose content in the massif is maximum. Conversely, the presence of uniformly distributed fragments in the massif poses the task of determining effective parameters of DBO, as the computed average value does not ensure adequate crushing of boulders. The energy spent on breakage would not be maximal and the seismic effect would increase.

If varying degrees of homogeneity exist in the quarry (Fig. 64,c), then the methods and values of parameters of DBO (means for achieving effective blast, assessed by physical indices) can be differentiated by the seismic effects, degree and evenness of breakage. In practice, the norms for specific explosive consumption should be differentiated taking into account the degree of homogeneity of massifs based on distribution of particle sizes.

Thus the nature of particle size distribution in the massif is the major influencing factor in the effectiveness of breakage by blasting. It can be used for controlling the quality of blasting in a selective manner. A blasting method, to achieve the required degree of fragmentation and qualitative indexes with minimum explosive consumption in the entire quarry, can be chosen along the zones of homogeneity.

The suggested evaluation of effectivity of the breakage method and complexity of the problem can be used in methodologies for computing blasting parameters and the physical effectiveness can be assessed both from the point of view of the powder factor and blast-induced seismic effects.

Literature Cited

1. Avarbukh, A.G. 1982. Izuchenie sostava i svoistv gornykh porod pri seis-morazvedke (Study of Composition and Properties of Rocks in Seismic Exploration). Nedra, Moscow.

2. Aku, K. and P. Richards. 1983. Kolichesvennaya seismologiya (Qualitative Seismology). Mir, Moscow, vol. 1.

3. Boiko, V.V. 1982. Ob opredelenii granits seismobezopasnoi zony v ani-zotropnykh massivakh (Determining boundaries of seismic safe zone in anisotropic massifs). In: *Deistvie Vzryva v Gruntakh i Gornykh Porodakh*, pp. 164–166. Naukova Dumka, Kiev.

4. Vorob'ev, V.D. 1982. Uchet anizotropii massiva pri vybore seismobezopas-nykh parametrov vzryvnykh rabot (Considering the anisotropy of a massif while selecting seismic safe parameters of blasting operations). *Dobycha Uglya Otkrytym Sposobom*, no. 5, pp. 6–7.

5. Vorob'ev, V.D. and I.F. Gonchar. 1981. O proektirovanii effektivnykh i bezoposnykh parametrov vzryvnykh rabot (On designing effective and safe blasting parameters). *Bezopasnost' Truda v Promyshlennosti*, no. 12, pp. 34–36.

6. Vorob'ev, V.D. and V.V. Peregudov. 1984. Vzryvnye raboty v skal'nykh porodakh (Blasting Operations in Hard Rocks). Naukova Dumka, Kiev.

7. Vorob'ev, V.D., A.A. Kuz'menko, S.N. Markelov and Yu.S. Sikorskii. 1986. Vliyanie vremeni zamedleniya vzryvom na seismicheskoe deistvie (Effect of delay interval on seismic effect). *Stroitel'nye Materialy i Konstruktsii*, no. 2, p. 37.

8. Vovk, A.A., V.G. Kravets, I.A. Luchko and A.V. Mikhalyuk. 1981. Geo-dinamika vzryva i ee prilozheniya (Geodynamics of Blasting and Its Appli-cation). Naukova Dumka, Kiev.

9. Gribanova, L.L., V.V. Kudinov and V.M. Tkachenko. 1982. Vliyanie glu-biny raspolozheniya vzryvamogo bloka v kar'ere na intensivnost' seismich-eskikh voln (The influence of depth of block being blasted in an open pit on the intensity of seismic waves). In: *Vzryvnoe Delo*, no. 85/42, pp. 127–132.

10. Dauetas, A.A., A.A. Kuz'menko and L.A. Furman. 1978. Vliyanie para-metrov vzryvnykh rabot na intensivnost' seismicheskikh kolebanii (Influence of blasting parameters on intensity of seismic vibrations). In: *Ispol'zovanie Vzryva pri Razrabotke Neskal'nykh Gruntov*, pp. 68–72. Naukova Dumka, Kiev.

11. Denisyuk, I.I. and A.S. Marchenko. 1984. Opredelenie napryazhennogo sostoyaniya gruntovogo massiva pri vzryve gruppy tsilindricheskikh zaryadov (Determination of stress state in a soil massif for a blast of a group of cylindrical charges). In: *Vzryvnye Raboty v Gruntakh i Gornykh Porodakh*, pp. 43–47, Naukova Dumka, Kiev.

12. Denisyuk, I.I. and A. Zakirov. 1982. Snizhenie seismicheskogo effekta pri vzryvnom uplotnenii strukturno-neustoichivykh gruntov (Reduction of blast-induced seismic effects in compacting structurally unstable soils). *Proektirovanie i Stroitel'stvo Truboprovodov i Gazoneftepromyslovykh Sooruzhenii*, no. 4, pp. 44–47.

13. Denisyuk, I.I., V.G. Kravets and A.S. Marchenko. 1985. Issledovanie spektra seismicheskikh kolebanii pri vzryve sfericheskogo zaryada (Studying seismic wave spectra in a spherical charge blast). In: *Vzryv v Gruntakh i Gornykh Porodakh*, pp. 78–84, Naukova Dumka, Kiev.

14. Dolgov, K.A. 1979. Opredelenie ekonomicheskoi effektivnosti primeneniya novykh tipov VV pri droblenii krepkikh gornykh porod vzryvom (Establishing the economic effectiveness of using new types of explosives in hard rock breakage by blasting). *Izv. Vuzov. Gornyi Zhurnal*, no. 4, pp. 67–69.

15. Drogoveiko, I.Z. 1981. Razrushenie merzlykh gruntov vzryvom (Breakage of Permafrost Soils by Blasting). Nedra, Moscow.

16. Kazakevich, L.D. 1979. Narodnokhozyaistvennye zatraty v planirovanii effektivnosti proizvodstva (Plan Outlay in Planning Effectiveness in Industry). Ekonomika, Moscow.

17. Korchinskii, I.L. 1971. Seismostoikoe stroitel'stvo zdanii (Seismically Stable Building Construction). Vysshaya Shkola, Moscow.

18. Kravets, V.G., I.I. Denisyuk and A.A. Kuz'menko. 1979. Metody rascheta i stroitel'stva pregrad s ispol'zovaniem energii vzryva (Methods of calculation and construction of screens using blast energy). In: *Vzryvnye Raboty v Gruntakh*, pp. 168–172. Naukova Dumka, Kiev.

19. Kudinov, V.V., L.P. Gribanova and L.L. Zhukova. 1983. Metody otsenki vrednykh proyavlenii vzryva i ikh snizheniya (Methods of evaluating harmful effects of blasting and their reduction). In: *Razvitie Progressivnykh Metodov Razrabotki Mestorozhdenii Zheleznykh i Margantsevykh rud USSR i Primenenie ikh na Predpriyatiyakh Otrasli*, pp. 92–93. Nauchn.-Issled. Gornorud. In-t, Krivoi Rog.

20. Kuz'menko, A.A. and I.I. Denisyuk. 1976. Razrabotka seismobezopasnogo metoda vedeniya vzryvnykh rabot (Development of safe seismic blasting methods). In: *Ispol'zovanie Energii Vzryva na Ob"ektakh Irrigatsionnogo i Meliorativnogo Stroitel'stva v Gruntakh*, pp. 137–143. Naukova Dumka, Kiev.

21. Kuz'menko, A.A. 1979. Seismicheskii effekt vzryvov tsilindricheskikh zaryadov (Seismic effect induced by blasting of cylindrical charges). In: *Vzryvnoe Delo*, no. 81/38, pp. 180–195.

22. Medvedev, L.A. 1979. Ekonomicheskaya otsenka norm raskhoda VV pri vzryvakh na vybros v gruntakh (Economic assessment of explosive consumption norms in cast blasts in soils). In: *Upravlenie Deistviem Vzryva v Gruntakh i Gornykh Porodakh*, pp. 105–110. Naukova Dumka, Kiev.

23. Metodika (osnovnye polozheniya) opredeleniya ekonomicheskoi effektivnosti ispol'zovaniya v narodnom khozyaistve novoi tekhniki, izobretenii i ratsionalizatorskikh predlozhenii [Methodology (Basic Fundamentals) for Determining the Economic Effectiveness of Utilising New Techniques, Inventions and Rationalisations in the National Economy]. 1977. VINITI, Moscow.

24. Mosinets, V.N. 1976. Drobyashchee i seismicheskoe deistvie vzryva v gornykh porodakh (Crushing and Seismic Effect Caused by Blasting in Rocks). Nedra, Moscow.

25. Mosinets, V.N. and A.V. Abramov. 1982. Razrushenie treshchinovatykh i narushennykh gornykh podov (Breakage of Jointed and Distributed Rocks). Nedra, Moscow.

26. Mosinets, V.N. and V.F. Bogatskii. 1983. Osnovnye nauchno-tekhnicheskie problemy seismiki blizhnei zony (Principal scientific-technical problems of seismic effects in the near-field zone). In: *Vzryvnoe Delo*, no. 84/42, pp. 89–101.

27. Mosinets, V.N., E.A. Grigor'yants and A.I. Teterin. 1983. Osobennosti seismicheskogo deistviya vzryvov na kar'ere s myagkimi pokryvayushchimi porodami (Peculiarities of seismic effects induced by blasts in an open pit with a soft overburden). In: *Vzryvnoe Delo*, no. 84/42, pp. 137–150.

28. Mosinets, V.N. 1986. Sovremennoe sostoyanie i perspektivy razvitiya tekhnologii i metodov proizvodstva vzryvnykh rabot na kap"lerakh SSSR (Current status and perspectives of development of technology and methods of blasting operations in open pit mines of USSR). In: *Vzryvnoe Delo*, no. 89/46, pp. 100–109.

29. Oksanich, I.F. and P.S. Mironov. 1982. Zakonomernosti drobleniya gornykh porod vzryvom i prognozirovanie granulometricheskogo sostava (Patterns of Rock Breakage by Blasting and Prediction of Granulometric Composition). Nedra, Moscow.

30. Vorob'ev, V.D., A.A. Kuz'menko, V.V. Zakharov *et al.* 1985. Opredelenie seismobezopasnykh parametrov massovykh vzryvov (Determining safe seismic parameters for large-scale blasts). *Stroitel'nye Materialy i Konstruktsii*, no. 2, pp. 33–34.

31. Kravets, V.G., B.I. Il'yasov, I.I. Denisyuk *et al.* 1982. Rekomendatsii po primeneniyu seismobezopasnykh metodov uplotneniya prosadochnykh lessovykh gruntov energiei vzryva (Recommendations for Adopting Safe Seismic Methods for Compacting Subsided Loess Soils by Blast Energy). TurkmenNIINTI, Ashkabad.

32. Rzhevskii, V.V. and G.Ya. Novik. 1984. Osnovy fiziki gornykh porod (Fundamental Physics of Rocks). Nedra, Moscow.

33. Mel'nikov, N.V. and V.V. Rzhevskii and M.M. Protod''yakonov (eds.). 1975. Spravochnik (kadastr) fizicheskikh svoistv gornykh porod (Reference Book on Physical Properties of Rocks). Nedra, Moscow.

34. Shteinbakh, N.A., B.K. Malikov and T.Zh. Borubaeva. 1984. K otsenke seismocheskogo deistviya vzryva na vybros pri stroitel'stve kanalov (Evaluating cast blast-induced seismic effect on canal construction). In: *Deistvie Vzryva v Gruntakh i Gornykh Porodakh*, pp. 74–89. Frunze.

35. Shteinbakh, N.A. 1979. Prognoz seismicheskogo effekta kombaritinskogo vzryva (Predicting the seismic effect of the Kombaritinskii blast). *Tr. Frunzenskogo Politekhn. In-ta*, vol. 112, pp. 137–146.

Milton Keynes UK
Ingram Content Group UK Ltd.
UKHW031133141024
449569UK00006B/226